Cache Replacement Policies

Synthesis Lectures on Computer Architecture

Editors
Natalie Enright Jerger, *University of Toronto*
Margaret Martonosi, *Princeton University*

Founding Editor Emeritus
Mark D. Hill, *University of Wisconsin, Madison*

Synthesis Lectures on Computer Architecture publishes 50- to 100-page publications on topics pertaining to the science and art of designing, analyzing, selecting and interconnecting hardware components to create computers that meet functional, performance and cost goals. The scope will largely follow the purview of premier computer architecture conferences, such as ISCA, HPCA, MICRO, and ASPLOS.

Cache Replacement Policies
Akanksha Jain and Calvin Lin
2019

The Datacenter as a Computer: Designing Warehouw-Scale Machines, Third Edition
Luiz André Barroso, Urs Hölzle, and Parthasarathy Ranganathan
2018

Principles of Secure Processor Architecture Design
Jakub Szefer
2018

General-Purpose Graphics Processor Architectures
Tor M. Aamodt, Wilson Wai Lun Fung, and Timothy G. Rogers
2018

Compiling Algorithms for Heterogenous Systems
Steven Bell, Jing Pu, James Hegarty, and Mark Horowitz
2018

Architectural and Operating System Support for Virtual Memory
Abhishek Bhattacharjee and Daniel Lustig
2017

Deep Learning for Computer Architects
Brandon Reagen, Robert Adolf, Paul Whatmough, Gu-Yeon Wei, and David Brooks
2017

On-Chip Networks, Second Edition
Natalie Enright Jerger, Tushar Krishna, and Li-Shiuan Peh
2017

Space-Time Computing with Temporal Neural Networks
James E. Smith
2017

Hardware and Software Support for Virtualization
Edouard Bugnion, Jason Nieh, and Dan Tsafrir
2017

Datacenter Design and Management: A Computer Architect's Perspective
Benjamin C. Lee
2016

A Primer on Compression in the Memory Hierarchy
Somayeh Sardashti, Angelos Arelakis, Per Stenström, and David A. Wood
2015

Research Infrastructures for Hardware Accelerators
Yakun Sophia Shao and David Brooks
2015

Analyzing Analytics
Rajesh Bordawekar, Bob Blainey, and Ruchir Puri
2015

Customizable Computing
Yu-Ting Chen, Jason Cong, Michael Gill, Glenn Reinman, and Bingjun Xiao
2015

Die-stacking Architecture
Yuan Xie and Jishen Zhao
2015

Single-Instruction Multiple-Data Execution
Christopher J. Hughes
2015

v

Power-Efficient Computer Architectures: Recent Advances
Magnus Själander, Margaret Martonosi, and Stefanos Kaxiras
2014

FPGA-Accelerated Simulation of Computer Systems
Hari Angepat, Derek Chiou, Eric S. Chung, and James C. Hoe
2014

A Primer on Hardware Prefetching
Babak Falsafi and Thomas F. Wenisch
2014

On-Chip Photonic Interconnects: A Computer Architect's Perspective
Christopher J. Nitta, Matthew K. Farrens, and Venkatesh Akella
2013

Optimization and Mathematical Modeling in Computer Architecture
Tony Nowatzki, Michael Ferris, Karthikeyan Sankaralingam, Cristian Estan, Nilay Vaish, and
David Wood
2013

Security Basics for Computer Architects
Ruby B. Lee
2013

The Datacenter as a Computer: An Introduction to the Design of Warehouse-Scale
Machines, Second Edition
Luiz André Barroso, Jimmy Clidaras, and Urs Hölzle
2013

Shared-Memory Synchronization
Michael L. Scott
2013

Resilient Architecture Design for Voltage Variation
Vijay Janapa Reddi and Meeta Sharma Gupta
2013

Multithreading Architecture
Mario Nemirovsky and Dean M. Tullsen
2013

Performance Analysis and Tuning for General Purpose Graphics Processing Units
(GPGPU)
Hyesoon Kim, Richard Vuduc, Sara Baghsorkhi, Jee Choi, and Wen-mei Hwu
2012

Automatic Parallelization: An Overview of Fundamental Compiler Techniques
Samuel P. Midkiff
2012

Phase Change Memory: From Devices to Systems
Moinuddin K. Qureshi, Sudhanva Gurumurthi, and Bipin Rajendran
2011

Multi-Core Cache Hierarchies
Rajeev Balasubramonian, Norman P. Jouppi, and Naveen Muralimanohar
2011

A Primer on Memory Consistency and Cache Coherence
Daniel J. Sorin, Mark D. Hill, and David A. Wood
2011

Dynamic Binary Modification: Tools, Techniques, and Applications
Kim Hazelwood
2011

Quantum Computing for Computer Architects, Second Edition
Tzvetan S. Metodi, Arvin I. Faruque, and Frederic T. Chong
2011

High Performance Datacenter Networks: Architectures, Algorithms, and Opportunities
Dennis Abts and John Kim
2011

Processor Microarchitecture: An Implementation Perspective
Antonio González, Fernando Latorre, and Grigorios Magklis
2010

Transactional Memory, Second Edition
Tim Harris, James Larus, and Ravi Rajwar
2010

Computer Architecture Performance Evaluation Methods
Lieven Eeckhout
2010

Introduction to Reconfigurable Supercomputing
Marco Lanzagorta, Stephen Bique, and Robert Rosenberg
2009

On-Chip Networks
Natalie Enright Jerger and Li-Shiuan Peh
2009

The Memory System: You Can't Avoid It, You Can't Ignore It, You Can't Fake It
Bruce Jacob
2009

Fault Tolerant Computer Architecture
Daniel J. Sorin
2009

The Datacenter as a Computer: An Introduction to the Design of Warehouse-Scale
Machines
Luiz André Barroso and Urs Hölzle
2009

Computer Architecture Techniques for Power-Efficiency
Stefanos Kaxiras and Margaret Martonosi
2008

Chip Multiprocessor Architecture: Techniques to Improve Throughput and Latency
Kunle Olukotun, Lance Hammond, and James Laudon
2007

Transactional Memory
James R. Larus and Ravi Rajwar
2006

Quantum Computing for Computer Architects
Tzvetan S. Metodi and Frederic T. Chong
2006

Cache Replacement Policies
Akanksha Jain and Calvin Lin

ISBN: 978-3-031-00634-0 paperback
ISBN: 978-3-031-01762-9 ebook
ISBN: 978-3-031-00059-1 hardcover

DOI 10.1007/978-3-031-01762-9

A Publication in the Springer series
SYNTHESIS LECTURES ON ADVANCES IN AUTOMOTIVE TECHNOLOGY

Lecture #47
Series Editors: Natalie Enright Jerger, *University of Toronto*
                Margaret Martonosi, *Princeton University*
Founding Editor Emeritus: Mark D. Hill, *University of Wisconsin, Madison*
Series ISSN
Print 1935-3235   Electronic 1935-3243

Cache Replacement Policies

Akanksha Jain and Calvin Lin
The University of Texas at Austin

SYNTHESIS LECTURES ON COMPUTER ARCHITECTURE #47

ABSTRACT

This book summarizes the landscape of cache replacement policies for CPU data caches. The emphasis is on algorithmic issues, so the authors start by defining a taxonomy that places previous policies into two broad categories, which they refer to as coarse-grained and fine-grained policies. Each of these categories is then divided into three subcategories that describe different approaches to solving the cache replacement problem, along with summaries of significant work in each category. Richer factors, including solutions that optimize for metrics beyond cache miss rates, that are tailored to multi-core settings, that consider interactions with prefetchers, and that consider new memory technologies, are then explored. The book concludes by discussing trends and challenges for future work. This book, which assumes that readers will have a basic understanding of computer architecture and caches, will be useful to academics and practitioners across the field.

KEYWORDS

hardware caches, cache replacement policies, cache hit rate, reducing memory latency, reducing memory traffic

Contents

Preface xiii

Acknowledgments xv

1 **Introduction** 1

2 **A Taxonomy of Cache Replacement Policies** 3
    2.1   Coarse-Grained Policies 4
    2.2   Fine-Grained Policies 5
    2.3   Design Considerations 6

3 **Coarse-Grained Replacement Policies** 9
    3.1   Recency-Based Policies 9
        3.1.1 Variants of LRU 10
        3.1.2 Beyond LRU: Insertion and Promotion Policies 13
        3.1.3 Extended Lifetime Recency-Based Policies 17
    3.2   Frequency-Based Policies 19
    3.3   Hybrid Policies 21
        3.3.1 Adaptive Replacement Cache (ARC) 22
        3.3.2 Set Dueling 22

4 **Fine-Grained Replacement Policies** 25
    4.1   Reuse Distance Prediction Policies 26
        4.1.1 Expiration-Based Dead Block Predictors 26
        4.1.2 Reuse Distance Ordering 27
    4.2   Classification-Based Policies 27
        4.2.1 Sampling Based Dead Block Prediction (SDBP) 28
        4.2.2 Signature Based Hit Prediction (SHiP) 30
        4.2.3 Hawkeye 31
        4.2.4 Perceptron-Based Prediction 33
        4.2.5 Evicted Address Filter (EAF) 33
    4.3   Other Prediction Metrics 35
        4.3.1 Economic Value Added (EVA) 35

5   Richer Considerations **39**

   5.1   Cost-Aware Cache Replacement 39

      5.1.1   Memory-Level Parallelism (MLP) 40

   5.2   Criticality-Driven Cache Optimizations 43

      5.2.1   Critical Cache 43

      5.2.2   Criticality-Aware Multi-Level Cache Hierarchy 44

   5.3   Multi-Core-Aware Cache Management 45

      5.3.1   Cache Partitioning 45

      5.3.2   Shared-Cache-Aware Cache Replacement 47

   5.4   Prefetch-Aware Cache Replacement 48

      5.4.1   Cache Pollution 49

      5.4.2   Deprioritizing Prefetchable Lines 50

   5.5   Cache Architecture-Aware Cache Replacement 53

      5.5.1   Inclusion-Aware Cache Replacement 53

      5.5.2   Compression-Aware Cache Replacement 53

   5.6   New Technology Considerations 55

      5.6.1   NVM Caches 55

      5.6.2   DRAM Caches 56

6   Conclusions **59**

Bibliography **63**

Authors' Biographies **71**

Preface

We have written this book for those who wish to understand the state of the art in cache replacement policies, so our goals are to explore the solution space and to organize the different approaches that have been explored in the literature. In doing so, we also hope to identify trends and issues that will be important in the future. We have intentionally chosen to focus on algorithmic issues, and we focus on cache replacement policies for CPU data caches. While most of the research that we discuss is performed in the context of last-level caches, where the benefits of intelligent cache replacement are most pronounced, the general ideas often apply to other levels of the cache hierarchy. After a brief introduction, Chapter 2 starts by defining our 2-dimensional taxonomy of cache policies. The primary dimension describes the granularity of replacement decisions, while the second one describes the metric that is used to make the replacement decisions. Chapters 3 and 4 then use our taxonomy to describe existing replacement policies. Chapter 5 introduces other considerations that complicate cache replacement, including data prefetchers, shared caches, variable miss costs, compression, and new technologies. We conclude in Chapter 6 by using the results of the 2017 Cache Replacement Championship to encapsulate recent trends, before stepping back and taking stock of larger movements and challenges for future research. We assume that readers have a basic undergraduate understanding of computer architecture and caches, but this book will be useful for researchers at all levels.

Akanksha Jain and Calvin Lin
May 2019

Acknowledgments

We thank Aamer Jaleel and the anonymous reviewer for their valuable feedback. We also thank Margaret Martonosi and Michael Morgan for their encouragement in writing this book and for helpful guidance and support throughout this process. This effort was funded in part by NSF Grant CCF-1823546 and a gift from Intel Corporation through the NSF/Intel Partnership on Foundational Microarchitecture Research, and it was funded in part by a grant from Huawei Technologies.

Akanksha Jain and Calvin Lin
May 2019

CHAPTER 1

Introduction

For decades now, the latency of moving data has greatly exceeded the latency of executing an instruction, so caches, which both reduce memory latency and reduce memory traffic, are important components of all modern microprocessors. Because there is a general tradeoff between the size of a cache and its latency, most microprocessors maintain a hierarchy of caches, with smaller lower-latency caches being fed by larger higher-latency caches, which is eventually fed by DRAM. For each of these caches, effectiveness can be measured by its hit rate, which we define to be $\frac{s}{r}$, where s is the number of memory requests serviced by the cache and r is the total number of memory requests made to the cache.

There are several methods of improving a cache's hit rate. One method is to increase the size of the cache, typically at the expense of increased latency. A second method is to increase the associativity of the cache, which increases the number of possible cache locations to which a cache line can be mapped. At one extreme, a direct mapped cache (associativity of 1) maps each cache line to a single location in the cache. At the other extreme, a fully associative cache allows a cache line to be placed anywhere in the cache. Unfortunately, power consumption and hardware complexity both increase as we increase associativity. The third method, which is the subject of this book, is to choose a good cache replacement policy, which answers the question, "When a new line is to be inserted into the cache, which line should be evicted to make space for the new line?"

It may seem strange to write a book on cache replacement, since Lazslo Belady produced a provably optimal policy over five decades ago. But Belady's policy is unrealizable because it relies on future knowledge—it evicts the line that will be re-used furthest in the future. Thus, over the years, researchers have explored several different approaches to solving the cache replacement problem, typically relying on heuristics that consider frequency of access, recency of access, and more recently, prediction techniques.

Moreover, there is the question of how cache replacement policies relate to dead block predictors, which attempt to predict lines that will no longer be needed in the cache. We now know that the life cycle of a cache line has multiple decision points, beginning with the insertion of the line and progressing over time to the eventual eviction of the line, and we know that there are different techniques for performing actions at these different decision points. With this view, we argue that dead block predictors are a special case of cache replacement policies, and we find that the space of cache replacement policies is quite rich.

Finally, caches are ubiquitous in software systems as well. In fact, the first replacement policies were developed for the paging systems of operating systems, and while there are technological differences between software caches and hardware caches, we hope that some of the ideas in this book will prove useful for the developers of software caches, as well.

Scope   This book focuses on hardware cache replacement policies for CPU data caches. And while most of the research that we discuss is performed in the context of last-level caches, where the benefits of intelligent cache replacement are most pronounced, the general ideas often apply to other levels of the cache hierarchy.

Roadmap   We start in Chapter 2 by defining a 2-dimensional taxonomy. The primary dimension describes the granularity of replacement decisions. A second dimension describes the metric that is used to make the replacement decisions. Chapters 3 and 4 then use our taxonomy to describe existing replacement policies. Chapter 5 introduces other considerations that complicate cache replacement, including data prefetchers, shared caches, variable miss costs, compression, and new technologies. We conclude in Chapter 6 by using the results of the Cache Replacement Championship held in 2017 to encapsulate recent trends in cache replacement, before stepping back and taking stock of larger trends and challenges for future research.

CHAPTER 2

A Taxonomy of Cache Replacement Policies

To organize both this book and the many ideas that have been studied over several decades, we present a taxonomy of solutions to the cache replacement problem. Our taxonomy is built on the observation that cache replacement policies solve a prediction problem, where the goal is to predict whether any given line should be allowed to stay in cache. This decision is re-evaluated at multiple points in a cache line's *lifetime*, which begins when the line is inserted into the cache and ends when the line is evicted from the cache.

Therefore, in our taxonomy, we first divide cache replacement policies into two broad categories based on the *granularity* of their insertion decisions. Polices in the first category, which we refer to as *Coarse-Grained policies*, treat all lines identically when they are inserted into the cache and only differentiate among lines based on their their behavior while they reside in the cache. For example, as a line resides in the cache, its priority might be increased each time it is reused. By contrast, *Fine-Grained policies* distinguish among lines when they are inserted into the cache (in addition to observing their behavior while they reside in the cache). To make this distinction at the time of insertion, Fine-Grained policies typically rely on historical information about cache access behavior. For example, if a Fine-Grained policy has learned that a particular instruction loads lines that in the past tend to be evicted without being reused, it can insert that line with a low priority.

To better understand our taxonomy, it is useful to understand that replacement policies are typically implemented by associating a small amount of *replacement state* with each cache line (Figure 2.1):

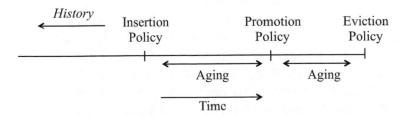

Figure 2.1: Operations of a replacement policy during a cache line's lifetime.

- *Insertion Policy*: How does the replacement policy initialize the replacement state of a new line when it is inserted into the cache?

- *Promotion Policy*: How does the replacement policy update the replacement state of a line when it hits in the cache?

- *Aging Policy*: How does the replacement policy update the replacement state of a line when a competing line is inserted or promoted?

- *Eviction Policy*: Given a fixed number of choices, which line does the replacement policy evict?

With this view of the lifetime of a cache line, we can see that *Coarse-Grained policies* treat all cache lines identically on insertion, so they primarily rely on clever aging and promotion policies to manipulate replacement priorities. By contrast, *Fine-Grained policies* employ smarter insertion policies. Of course, Fine-Grained policies also need appropriate aging and promotion to update replacement state because not all behavior can be predicted at the time of insertion.

We now discuss these two classes in more detail.

2.1   COARSE-GRAINED POLICIES

Coarse-Grained policies treat all lines identically when they are first inserted into the cache, and they distinguish cache-friendly lines from cache-averse lines by observing reuse behavior while the line resides in the cache. We further divide the Coarse-Grained policies into three classes based on the metric they use to differentiate cache-resident lines. The first class, which includes the vast majority of Coarse-Grained policies, describes policies that order cache lines based on their recency of access. The second class includes policies that order cache lines based on their frequency of access. Finally, policies in the third class monitor cache behavior over time to dynamically choose the best Coarse-Grained policy for a given time period.

Recency-Based Policies   Recency-Based Policies order lines based on each line's latest reference within its lifetime. To arrive at this ordering, Recency-Based policies maintain a conceptual *recency stack* that provides the relative order in which lines were referenced. Different policies exploit recency information in different ways. For example, the commonly used LRU (least recently used) policy (and its variants) preferentially evict least recently used lines, whereas policies such as MRU (most recently used) preferentially evict most recently used lines; other policies exploit intermediate solutions. We discuss such policies in Section 3.1.

We further divide Recency-Based Policies based on their definition of the lifetime during which recency behavior is observed. The first class, which includes the vast majority of Recency-Based policies, describes policies that end a line's lifetime when it is evicted from the cache. We consider such policies to have a *fixed lifetime*. The second class includes policies that extend a cache line's lifetime beyond eviction by introducing a temporary structure that gives lines with longer reuse distance a second chance to receive cache hits. We say that such policies have an *extended lifetime*.

Frequency-Based Policies   Frequency-Based Policies maintain *frequency counters* to order lines based on the frequency with which they are referenced. Different replacement policies use different strategies to update and interpret these frequency counters. For example, some policies increase the counters monotonically, while others age the counters (decrease their values) as time passes by. As another example, some policies evict lines with the lowest frequency, while others evict lines whose frequency satisfies some pre-defined criterion. We discuss representative solutions in Section 3.2.

Hybrid Policies   Since different Coarse-Grained policies work well for different cache access patterns, hybrid policies dynamically choose among a few pre-determined Coarse-Grained policies to adapt to phase changes in a program's execution. In particular, hybrid policies observe the hit rate of a few candidate Coarse-Grained policies during an evaluation period and use this feedback to choose the best Coarse-Grained policy for future accesses. Adaptive policies are advantageous because they help overcome pathologies of individual Coarse-Grained policies. In Chapter 3, we will see that state-of-the-art hybrid policies modulate between Coarse-Grained policies that use different insertion priorities, and we note that despite the change in insertion priority over time, such policies are still coarse-grained because within a time period, all lines are treated identically.

2.2   FINE-GRAINED POLICIES

Fine-Grained policies distinguish among lines when they are inserted into the cache. They make these distinctions by using information from a cache line's previous lifetimes. For example, if a line has in the past receive no hits, then it can be inserted with low priority. Since remembering the past behavior of *all* cache lines is infeasible, Fine-Grained policies typically remember caching behavior for *groups* of lines. For example, many policies combine into one group of cache lines that were last accessed by the same load instruction.

Of course, one consideration for all Fine-Grained policies is the metric that is used to distinguish these groups at the time of insertion. We divide Fine-Grained policies into two broad categories based on the metric used: (1) *Classification-Based Policies* associate with each group one of two possible predictions, namely, cache-friendly and cache-averse; and (2) *Reuse Distance-Based Policies* try to predict detailed reuse distance information for each group of cache lines. These two categories define extreme ends of a spectrum, where policies on one end of the spectrum have just two possible predicted values, and policies on the other end of the spectrum have many possible predicted values; many policies will lie in the middle by predicting one of, say, four or eight possible values. Finally, Beckmann and Sanchez propose novel metrics [Beckmann and Sanchez, 2017], which we discuss under a third category.

Classification-Based Policies   Classification-based policies classify incoming cache lines into two categories: cache-friendly lines or cache-averse. The main idea is to preferentially evict cache-averse lines to leave more space for cache-friendly lines, so cache-friendly lines are inserted with higher priority than cache-averse lines. A secondary ordering is maintained among cache-friendly lines, typically based on recency. Classification-based policies are widely regarded as state-of-the-art because (1) they can exploit a long history of past behavior to take smarter decisions for the future, and (2) they can accommodate all kinds of cache access patterns. We discuss these policies in Section 4.2.

Reuse Distance-Based Policies   Reuse Distance-Based Policies predict detailed reuse distance information for incoming lines. Lines that exceed the expected reuse distance without receiving hits are evicted from the cache. These policies can be viewed as predictive variants of either Recency-Based Policies or Frequency-Based Policies, because both Recency-Based and Frequency-Based policies implicitly estimate reuse distances (via recency and frequency, respectively) by monitoring a cache line's reuse behavior while it is cache-resident. The explicit prediction of reuse distances using historical information presents unique advantages and disadvantages, which we discuss in Section 4.1.

2.3   DESIGN CONSIDERATIONS

The primary goal of replacement policies is to increase cache hit rates, and many design factors contribute toward achieving a higher hit rate. We outline three such factors as follows.

- *Granularity*: At what granularity are lines distinguished at the time of insertion? Are all cache lines treated the same, or are they assigned different priorities based on historical information?

- *History*: How much historical information does the replacement policy utilize in making its decisions?

- *Access Patterns*: How specialized is the replacement policy to certain access patterns? Is it robust to changes in access patterns or to mixes of different access patterns?

Figure 2.2 summarizes our complete taxonomy and shows how different classes of replacement policies address these design factors. In general, the trend, as we move to the right of the figure, which generally corresponds with newer policies, is toward finer-grained solutions that use longer histories and that can accommodate a wide variety of access patterns. The importance of predicting cache behavior at fine granularity can be gauged by observing that the top four solutions in the recent Cache Replacement Championship (2017) were all fine-grained. Fine-grained predictions give state-of-the-art Fine-Grained policies two advantages. First, they allow Fine-Grained policies to only dedicate cache space to lines that are most likely to benefit from caching; by contrast, Coarse-Grained policies tend to repeatedly dedicate cache resources to lines that do not yield any hits. Second, they allow Fine-Grained policies to dynamically determine access patterns for different groups of lines; by contrast, Coarse-Grained policies assume that the entire cache follows a uniform cache access pattern.

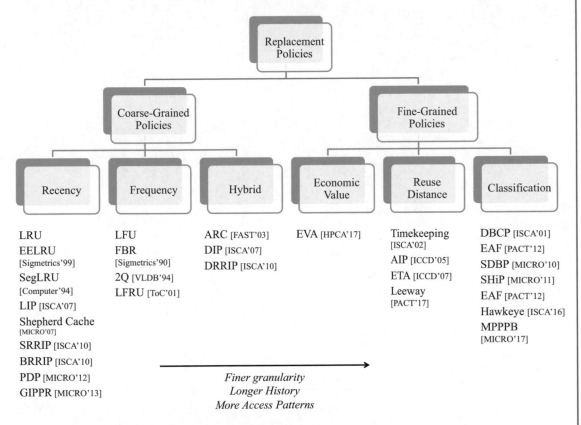

Figure 2.2: A taxonomy of cache replacement policies.

Among Fine-Grained policies, the trend is toward using longer histories as the state-of-the-art Fine-Grained policy [Jain and Lin, 2016] samples information from a history that is 8× the size of the cache (see Chapter 4). The use of long histories allows the policy to detect cache-friendly lines that have long-term reuse, which would otherwise be obfuscated by a long sequence of intermediate cache-averse accesses.

CHAPTER 3

Coarse-Grained Replacement Policies

The cache replacement literature spans 50 years, first in the context of OS page replacement and then in the context of hardware caches. During this time, most research has focused on the development of smarter Coarse-Grained policies, which can perhaps be explained by the simplicity of these policies: each cache line is associated with a small amount of replacement state, which is initialized uniformly for all newly inserted lines and then manipulated using simple operations as the lines are reused.

In this chapter, we discuss key advances in Coarse-Grained policies. We divide Coarse-Grained policies into three classes based on the manner in which they distinguish lines after their insertion into the cache. The first class, which includes the vast majority of Coarse-Grained policies, uses recency information to order lines (Recency-Based Policies). The second class instead uses frequency to order lines (Frequency-Based Policies). Finally, the third class (Hybrid polices) dynamically chooses among different Coarse-Grained replacement policies.

One running theme in the design of Coarse-Grained policies is the recognition of three commonly observed cache access patterns, namely, *recency-friendly accesses*, *thrashing accesses* [Denning, 1968], and *scans* [Jaleel et al., 2010b]. Recency-friendly accesses occur when the working set of the application is small enough to fit in the cache, such that the reuse distance of most memory references is smaller than the size of the cache. By contrast, thrashing accesses occur when the working set of the application exceeds the size of the cache, such that the reuse distance of most memory references is greater than the size of the cache. Finally, scans are a sequence of streaming accesses that never repeat. As we will see in this chapter, almost all Coarse-Grained replacement policies are optimized specifically for these access patterns or for mixes of these three access patterns.

3.1 RECENCY-BASED POLICIES

Recency-based policies prioritize lines based on recency information. The LRU policy [Mattson et al., 1970] is the simplest and most widely used of these policies. On a cache eviction, the LRU policy simply evicts the oldest line among a set of given candidates. To find the oldest line, the LRU policy conceptually maintains a recency stack, where the top of the stack represents the MRU line, and the bottom of the stack represents the LRU line. This recency stack can be maintained either precisely or approximately [Handy, 1998] by associating with each line a counter and updating it, as shown in Table 3.1.

Table 3.1: The LRU policy

Insertion	Promotion	Aging	Victim Selection
MRU position	MRU position	Move 1 position toward LRU	LRU position

LRU performs well when there is temporal locality of reference, that is, when data that was used recently is likely to be reused in the near future. But it performs poorly for two types of access patterns. First, it can be pessimal when the application's working set size exceeds the cache size, leading to a phenomenon known as *thrashing* [Denning, 1968] (see Figure 3.1). Second, LRU performs poorly in the presence of scans because it caches more recently used scans at the expense of older lines that are more likely to be reused.

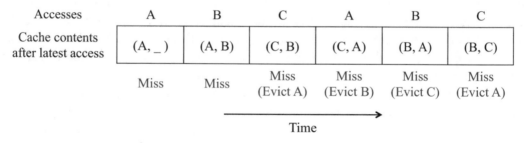

Figure 3.1: An example of thrashing. The LRU policy produces no cache hits when the access pattern cycles through a working set (3 in this example) that is larger than the cache capacity (2 in this case).

3.1.1 VARIANTS OF LRU

We now describe some of the notable variants of LRU that detect and accommodate thrashing or streaming accesses.

Most Recently Used (MRU)   The MRU Policy addresses thrashing by evicting new lines to retain old lines. Thus, when the working set is larger than the cache, it is able to retain a portion of the working set. For example, Figure 3.2 shows that for the thrashing access pattern shown in Figure 3.1, MRU improves upon LRU's hit rate by caching a portion of the working set—in this example never evicting line *A*. Table 3.2 shows that the MRU policy is identical to LRU, except that it evicts the line at the MRU position instead of the LRU position.

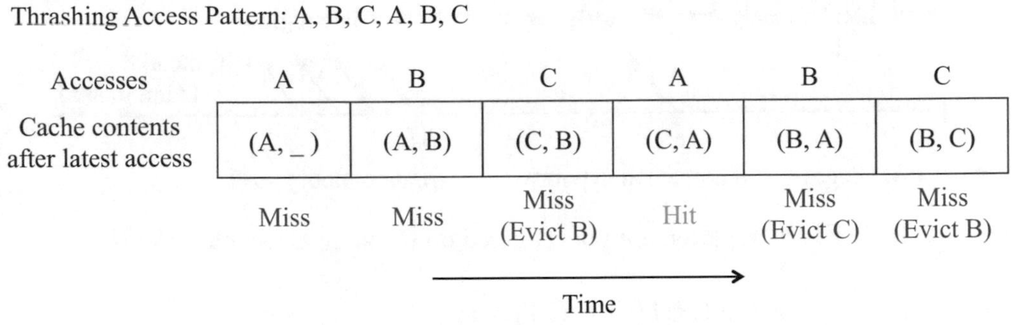

Thrashing Access Pattern: A, B, C, A, B, C

Accesses	A	B	C	A	B	C
Cache contents after latest access	(A, _)	(A, B)	(C, B)	(C, A)	(B, A)	(B, C)
	Miss	Miss	Miss (Evict B)	Hit	Miss (Evict C)	Miss (Evict B)

Time →

Figure 3.2: The MRU policy avoids thrashing by caching a portion of the working set. In this example, the MRU policy is able to cache A even though the working set size exceeds the cache capacity of two lines.

Table 3.2: The MRU policy

Insertion	Promotion	Aging	Victim Selection
MRU position	MRU position	Move 1 position toward LRU	MRU position

While MRU is ideal for thrashing accesses, it performs poorly in the presence of recency-friendly accesses, and it adapts poorly to changes in the application's working set, as it is unlikely to cache anything from the new working set.

Early Eviction LRU (EELRU)   The EELRU policy also avoids thrashing [Smaragdakis et al., 1999]. The main idea is to detect cases where the working set size exceeds the cache size, at which point a few lines are evicted early. Thus, early eviction discards a few randomly chosen lines so that the remaining lines can be managed effectively by the LRU policy. More specifically, EELRU evicts the LRU line when the working set fits in the cache, but it evicts the eth most recently used line when it observes that too many lines are being touched in a roughly cyclic pattern that is larger than main memory.

Figure 3.3 shows that EELRU distinguishes among three regions of the recency axis. The left endpoint of the recency axis represents the MRU line, and the right endpoint represents the LRU line. The LRU memory region consists of the most recently used lines, and positions e and M mark the beginning of the *early eviction region* and *late eviction region*, respectively. On a miss, the EELRU policy either evicts the line at the least recently used position (late region) or the line at the eth position (early region).

To decide whether to use early eviction or LRU eviction, EELRU tracks the number of hits received in each region. If the distribution is monotonically decreasing, then EELRU assumes that there is no thrashing and evicts the LRU line. But if the distribution shows more hits

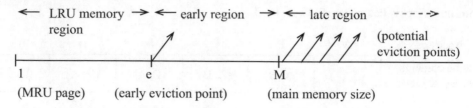

Figure 3.3: The EELRU replacement policy; based on [Smaragdakis et al., 1999].

Table 3.3: The early eviction LRU (EELRU) policy

Insertion	Promotion	Aging	Victim Selection
MRU position	MRU position	Move 1 position toward LRU	Choose between LRU or e^{th} recently used line

in the late region than the early region, then EELRU evicts from the early region, which allows lines from the late region to remain longer in the cache. Table 3.3 summarizes the operation of EELRU.

Segmented LRU (Seg-LRU)   Segmented LRU handles scanning accesses by preferentially retaining lines that have been accessed at least twice [Karedla et al., 1994]. Seg-LRU divides the LRU stack into two logical segments (see Figure 3.4), namely, the *probationary segment* and the *protected segment*. Incoming lines are added to the probationary segment and are promoted to the protected segment when they receive a hit. Thus, lines in the protected segment have been accessed at least twice, and scanning accesses are never promoted to the protected segment. On an eviction, the least recently used line from the probationary segment is selected.

Table 3.4 summarizes the operation of Seg-LRU. New lines are inserted at the MRU position in the probationary segment, and on a cache hit, lines are promoted to the MRU position in the protected segment. Because the protected segment is finite, a promotion to the protected segment may force the migration of the LRU line in the protected segment to the MRU end of the probationary segment, giving this line another chance to receive a hit before it is evicted from the probationary segment. Thus, Seg-LRU can adapt to changes in the working set as old lines eventually get demoted to the probationary segment.

Table 3.4: The Seg-LRU policy

Insertion	Promotion	Aging	Victim Selection
MRU position in probationary segment	MRU position in protected segment	Increment recency counter	LRU position from probationary segment

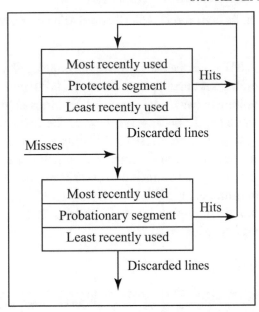

Figure 3.4: The Seg-LRU replacement policy; based on [Karedla et al., 1994].

A variant of the Seg-LRU policy won the First Cache Replacement Championship [Gao and Wilkerson, 2010].

3.1.2 BEYOND LRU: INSERTION AND PROMOTION POLICIES

Qureshi et al. [2007] observe that variants of Recency-Based policies can be realized by modifying the insertion policy, while keeping the eviction policy unchanged (evict the line in the LRU position). For example, MRU (Table 3.2) can be emulated by using the *LRU Insertion policy (LIP)* [Qureshi et al., 2007], which inserts new lines in the LRU position instead of the MRU position (see Table 3.5).

Table 3.5: The LRU insertion policy (LIP) emulates MRU

Insertion	Promotion	Aging	Victim Selection
LRU position	MRU position	Move 1 position toward LRU	LRU position

This insight spurred the design of replacement policies with new insertion and promotion policies. By interpreting recency information in different ways, these policies can address much broader classes of access patterns than LRU. In this section, we discuss some notable solutions in this direction. Since applications typically consist of many different access patterns, none of these policies is sufficient on its own and is typically used as part of a hybrid solution, which we discuss in Section 3.3.

Bimodal Insertion Policy (BIP)   Since thrash-resistant policies like LIP (and MRU) cannot adapt to changes in the working set during phase changes, BIP [Qureshi et al., 2007] modifies LIP such that lines are occasionally (with a low probability) inserted in the MRU position. BIP maintains the thrash-resistance of LIP because it inserts in the LRU position most of time, but it can also adapt to phase changes by occasionally retaining newer lines. Table 3.6 shows that BIP inserts in the MRU position with a probability of ϵ, which is set to be 1/32 or 1/64. An ϵ value of 1 inserts in the MRU position (mimicking LRU's insertion policy), and an ϵ value of 0 inserts in the LRU position (mimicking LIP's insertion policy). Thus, from an implementation point of view, BIP's ϵ parameter unifies all insertion policies that lie at different points in the spectrum between LRU and MRU insertion.

Table 3.6: The bimodal insertion policy (BIP)

Insertion	Promotion	Aging	Victim Selection
MRU position with probability ϵ	MRU position	Move 1 position toward LRU	LRU position

Static Re-Reference Interval Prediction (SRRIP)   Jaleel et al. [2010b] observe that cache replacement can be thought of as a *Re-Reference Interval Prediction (RRIP)* problem, and the conventional LRU chain can be instead thought of as an RRIP chain; while a line's position in the LRU chain represents the amount of time since its last use, a line's position in the RRIP chain represents the order in which it is predicted to be re-referenced [Jaleel et al., 2010b]. In particular, the line at the head of the RRIP chain is predicted to be re-referenced the soonest (shortest re-reference interval), while the line at the tail of the RRIP chain is predicted to be reused furthest in the future (largest re-reference interval). On a cache miss, the line at the tail of the RRIP chain is replaced.

   With this view, we can see that the LRU policy predicts that new lines will have a near-immediate re-reference interval (insert at the head of the RRIP chain), while the thrash-resistant LIP policy predicts that new lines will have a distant re-reference interval (insert at the tail of the RRIP chain). Figure 3.5 illustrates this point with an RRIP chain that is represented using a 2-bit Re-Reference Prediction Value (RRPV): 00 corresponds to the nearest re-reference interval prediction, 11 corresponds to a distant re-reference interval prediction.

   Jaleel et al. [2010b] show that instead of predicting re-reference intervals that lie at the extreme ends of the RRIP chain, there is great benefit in predicting an *intermediate* re-reference interval, which allows the replacement policy to accommodate a mixture of different access patterns. In particular, *scans*—a burst of references to data that is not reused—disrupt recency-

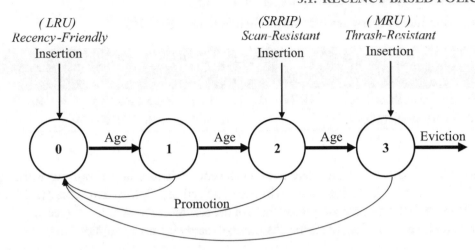

Figure 3.5: An example RRIP chain with 2-bit RRPV values.

friendly policies, such as LRU, because they discard the working set of the application without yielding any cache hits. To accommodate mixes of recency-friendly accesses and scans, Jaleel et al. [2010b] propose SRRIP, which gives incoming lines an intermediate re-reference interval and then promotes lines to the head of the chain if they are reused. Thus, SRRIP adds scan-resistance to recency-friendly policies, as it prevents lines with a distant re-reference interval (scans) from evicting lines with a near re-reference interval.

In the most general case, SRRIP associates an M-bit value per cache block to store its Re-Reference Prediction Value (RRPV), but Jaleel et al. [2010b] find that a 2-bit RRPV value is sufficient to provide scan-resistance. Table 3.7 summarizes the operations of SRRIP with a 2-bit RRPV value.

Table 3.7: Static re-reference interval prediction policy (SRRIP)

Insertion	Promotion	Aging	Victim Selection
RRPV = 2	RRPV = 0	Increment all RRPVs (if no line with RRPV = 3)	RRPV = 3

Like LRU, SRRIP thrashes the cache when the re-reference interval of all blocks is larger than the available cache. To add thrash-resistance to scan-resistant policies, Jaleel et al. [2010b] propose *Bimodal RRIP (BRRIP)*, a variant of BIP [Qureshi et al., 2007] that inserts a majority of cache blocks with a distant re-reference interval prediction (i.e., RRPV of 3), and it infrequently inserts cache blocks with an intermediate re-reference interval prediction (i.e., RRPV of 2). BRRIP's operations are summarized in Table 3.8.

Table 3.8: Bimodal re-reference interval prediction policy (BRRIP)

Insertion	Promotion	Aging	Victim Selection
RRPV=3 for most insertions, RRPV=2 with probability ϵ	RRPV = 0	Increment all RRPVs (if no line with RRPV = 3)	RRPV = 3

Since applications can alternate between recency-friendly and thrashing working sets, neither SRRIP nor BRRIP is sufficient on its own. In Section 3.3, we will discuss hyrbid versions of SRRIP and BRRIP that can address all three of the common access patterns (recency-friendly, thrashing and scans), yielding performance improvements for many applications.

Protecting Distance-Based Policy (PDP)   The PDP policy [Duong et al., 2012] is a generalization of RRIP as it dynamically estimates a Protecting Distance (PD), and all cache lines are protected for PD accesses before they can be evicted. The PD is a single value that is used for all lines inserted into the cache, but it is continually updated based on the application's dynamic behavior.

To estimate the PD, PDP computes the reuse distance distribution (RDD), which is the distribution of reuse distances observed within a recent interval of the program's execution. Using the RDD, the PD is defined to be the reuse distance that covers a majority of lines in the cache, such that most lines are reused at the PD or smaller distance. For example, Figure 3.6 shows the RDD for 436.cactusADM, where the PD is set to be 64. The PD is recomputed infrequently using a small special-purpose processor.

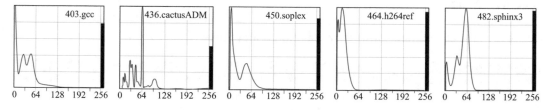

Figure 3.6: Protecting distance covers a majority of lines (for example, 64 for 436.cactusADM); based on [Duong et al., 2012].

More concretely, on an insertion or promotion, each cache line is assigned a value to represent its remaining PD (RPD), which is the number of accesses for which it remains protected; this value is initialized to the PD. After each access to a set, the RPD of each line in the set is aged by decrementing the RPD value (saturating at 0). A line is protected only if its RPD is larger than 0. An unprotected line is chosen as the victim.

Genetic Insertion and Promotion for PseudoLRU Replacement (GIPPR)   Taking inspiration from RRIP, Jiménez [2013] observe that there are many degrees of freedom in the choice of insertion and promotion, so they generalize modifications to insertion and promotion policies using the concept of an *Insertion/Promotion Vector (IPV)*. The IPV specifies a block's new position in the recency stack both when it is inserted into the cache and when it is promoted from a different position in the recency stack. In particular, for a k-way set-associative cache, an IPV is a $k + 1$-entry vector of integers ranging from 0 to $k - 1$, where the value in the ith position represents the new position to which a block in position i should be promoted when it is re-referenced. The kth entry in the IPV represents the position where a new incoming block should be inserted. If the new position in the recency stack is less than the old position, then blocks are shifted down to make room; otherwise blocks are shifted up to make room.

Figure 3.7 shows two sample IPVs, with the first one representing LRU, and the second one representing a more sophisticated insertion and promotion strategy.

While the generalized notion of IPVs is quite powerful, the number of possible IPVs grows exponentially, (k^{k+1} possible IPVs for a k-way cache), so Jiménez [2013] use an offline genetic search to evolve good IPVs for the SPEC 2006 benchmarks. The genetic search yielded the IPV shown in the bottom part of Figure 3.7.

Just as there is no single insertion policy or RRIP policy that works for all workloads, the best IPV differs for each workload. Therefore, Jiménez [2013] present a hybrid solution that uses set dueling (described in Section 3.3) to consider multiple IPVs.

3.1.3   EXTENDED LIFETIME RECENCY-BASED POLICIES

Extended Lifetime Policies are a special class of Recency-Based policies that artificially extend the lifetimes of some cache lines by storing them in an auxiliary buffer or a *victim cache*. The key motivation here is to defer eviction decisions to a later time when a more informed decision can be made. This strategy allows Coarse-Grained policies to slightly expand the reuse window of cache hits to be larger than the size of the cache.

Shepherd Cache   The Shepherd Cache (SC) mimics Belady's optimal policy [Rajan and Govindarajan, 2007]. Since Belady's policy requires knowledge of future accesses, Rajan and Govindarajan emulate this future lookahead with the help of an auxiliary cache, called the Shepherd Cache. In particular, the cache is logically divided into two components, the Main Cache (MC) which emulates optimal replacement, and the SC which uses a simple FIFO replacement policy. The SC supports optimal replacement for the MC by providing a lookahead window. New lines are initially buffered in the SC, and the decision to replace a candidate from the MC is deferred until the new line leaves the SC. While the new line is in the SC, information is gathered about the reuse order of replacement candidates in the MC. For example, candidates that are reused earlier become less likely candidates for eviction since Belady's policy evicts the lines that is reused furthest in the future. When the new line is evicted from the SC (due to other insertions in the SC), a replacement candidate is chosen by either picking a candidate from the

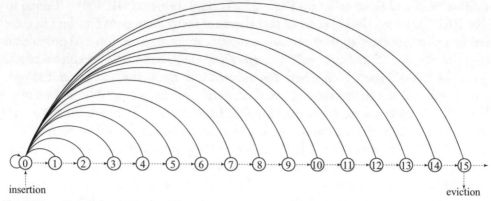

(a) Transition Graph for LRU. A solid edge points to the new position for an accessed or inserted block. A dashed edge shows where a block is shifted when another block is moved to its position.

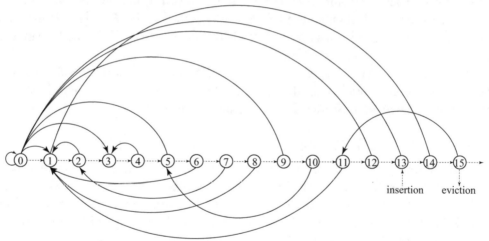

(b) Transition Graph for Vector [0 0 1 0 3 0 1 2 1 0 5 1 0 0 1 11 13].

Figure 3.7: IPV for LRU (top), and the IPV evolved using genetic algorithm (bottom); based on [Jiménez, 2013].

MC that hasn't been reused within the lookahead window, or the candidate that was reused last; if all lines in the MC were reused before the SC line was reused, then the SC line replaces itself. Although SC and MC are logically separate, Rajan and Govindarajan [2007] avoid any movement of data from one component to another by organizing the cache such that the two logical components can be organized as a single physical structure.

Thus, the SC emulates a superior replacement scheme for lines in the MC cache by extending their lifetime with the help of the SC. The tradeoff for SC is that replacement in the MC approaches true optimality with high lookaheads, and the higher lookahead comes at the cost of a diminished MC capacity. Unfortunately, subsequent work by Jain and Lin [2016] has shown that to approach the behavior of Belady's optimal policy, the policy needs a lookahead of 8× the size of the cache.

3.2   FREQUENCY-BASED POLICIES

Instead of relying on recency, Frequency-Based policies use access frequency to identify victims, so lines that are accessed more frequently are preferentially cached over lines that are accessed less frequently. This approach is less susceptible to interference from scans and has the benefit of accounting for reuse behavior over a longer period of time, as opposed to just the last use.

The simplest Frequency-Based policy is the Least Frequently Used (LFU) policy [Coffman and Denning, 1973], which associates a frequency counter with each cache line. The frequency counter is initialized to 0 when the line is inserted into the cache, and it is incremented each time the line is accessed. On a cache miss, the candidate with the lowest frequency is evicted. Table 3.9 summarizes these operations.

Table 3.9: Least frequently used (LFU) policy

Insertion	Promotion	Aging	Victim Selection
Frequency = 0	Increment Frequency	n/a	Least Frequency

Unfortunately, Frequency-Based policies adapt poorly to changes in the application's phases because lines with high frequency counts from a previous phase remain cached in new phases even when they are no longer being accessed. To address this issue, several solutions [Lee et al., 2001, O'Neil et al., 1993, Robinson and Devarakonda, 1990, Shasha and Johnson, 1994] augment frequency information with recency information to allow old lines to age gracefully. Here we discuss two such solutions.

Frequency-Based Replacement (FBR)   FBR [Robinson and Devarakonda, 1990] notes that Frequency-Based methods are susceptible to misleadingly high counter values from short bursts of temporal locality. Therefore, FBR factors out locality from frequency counts by selectively incrementing frequency counters. In particular, FBR does not increment frequency counters for the top portion of the LRU stack, which is known as the *new section*. Thus, short bursts of temporal locality do not affect the frequency counters. Figure 3.8 illustrates this strategy.

Unfortunately, this strategy has the disadvantage that once lines age out of the *new section*, even frequently used lines are evicted quickly because they do not have enough time to build up their frequency counts. Therefore, FBR further restricts replacement to the least frequently used line in an *old section*, which consists of lines that have not been recently accessed (bottom of the LRU stack). The remaining part of the stack is called the *middle section*, which gives frequently used lines sufficient time to build up their frequency values.

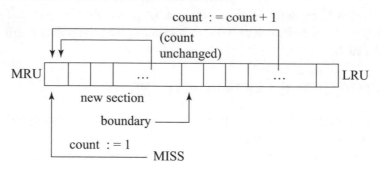

Figure 3.8: FBR does not increment frequency counters in the new section; based on [Robinson and Devarakonda, 1990].

Table 3.10 summarizes the operation of FBR policy.

Table 3.10: The frequency-based replacement (FBR) policy

Insertion	Promotion	Aging	Victim Selection
MRU position Frequency = 0	MRU position Frequency++ if not in new section	Increment by 1	LFU line in old section

Least Recently/Frequently Used (LRFU)   The LRFU policy [Lee et al., 2001] builds on the observation that the LRU and LFU policies represent extreme points of a spectrum of policies that combine recency and frequency information (see Figure 3.9). Using a new metric called Combined Recency and Frequency (CRF), LRFU explores this spectrum by allowing a flexible tradeoff between recency and frequency.

Like Frequency-Based policies, LRFU accounts for each past reference to the block, but unlike Frequency-Based policies, LRFU weighs the relative contribution of each reference by a *weighing function*. In particular, LRFU computes for each block a CRF value, which is the sum of the weighing function $F(x)$ for each past reference, where x is the distance of the past reference from the current time. Thus, for emulating purely Frequency-Based policies, the weighing function can give equal priority to all past references, and for emulating Recency-Based policies, the weighing function can give high priority to the last reference of the line.

LRFU uses the weighing function in Equation 3.1, where λ is an empirically chosen parameter. The weighing function gives exponentially lower priority to older lines, which allows LRFU to retain the benefits of FBR, while supporting graceful aging.

$$F(x) = \left(\tfrac{1}{p}\right)^{\lambda x}. \tag{3.1}$$

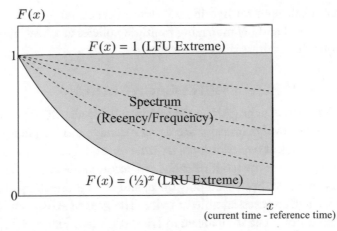

Spectrum of LRFU according to the function $F(x) = (\frac{1}{2})^{\lambda x}$,
where x is (current_time − reference_time).

Figure 3.9: The LRFU replacement policy; based on [Lee et al., 2001].

Table 3.11 summarizes the operation of LRFU policy for a block b at different decision points.

Table 3.11: The least recently/frequently used (LRFU) policy

Insertion	Promotion	Aging	Victim Selection
$CRF(b) = F(0)$ $LAST(b) = t_c$	$CRF(b) = F(0) +$ $F(t_c - LAST(b)) * CRF_{last}(b)$ $LAST(b) = t_c$	$t_c = t_c + 1$	Line with min CRF value

The performance of LRFU heavily depends on λ, so the subsequently developed ALRFU policy dynamically adjusts the value of lambda [Lee et al., 1999].

3.3 HYBRID POLICIES

Hybrid policies [Jaleel et al., 2010b, Qureshi et al., 2006, 2007] recognize that different workloads, or even different phases within the same workload, benefit from different replacement strategies. For example, if a program alternates between small and large working sets, it will benefit from a recency-friendly policy when the working set it small and from a thrash-resistant policy when the working set is large. Thus, hybrid policies assess the requirements of the application's current working set and dynamically choose from among multiple competing policies.

The two main challenges for hybrid policies are (1) accurately identifying the policy that will be the most beneficial and (2) managing multiple policies at a low hardware cost. We now discuss two solutions that address these challenges.

3.3.1   ADAPTIVE REPLACEMENT CACHE (ARC)

The Adaptive Replacement Cache (ARC) [Megiddo and Modha, 2003] combines the strengths of recency and frequency by dynamically balancing recency- and frequency-based evictions. In particular, ARC keeps track of two additional tag directories, L1 and L2, which together remember twice as many elements as the baseline cache can accommodate. The L1 directory maintains recently used pages that have been seen only once, and the L2 directory maintains recently accessed pages that have been accessed at least twice. The goal of ARC is to dynamically choose the appropriate amount of cache to dedicate to L1 and L2 (see Figure 3.10).

More concretely, ARC splits the baseline cache directory into two lists, T1 and T2, for recently and frequently referenced entries, respectively. T1 and T2 are extended with ghost lists (B1 and B2, respectively), which track recently evicted cache entries from T1 and T2, respectively. Ghost lists only contain tag metadata, not the actual data, and entries are added to ghost lists when the corresponding data is evicted from the cache. T1 and B1 together form the Recency-Based L1 directory, and T2 and B2 together form the Frequency-Based L2 directory.

ARC dynamically modulates the cache space dedicated to T1 and T2. In general, hits in B1 increase the size of T1 (proportion of cache dedicated to recently-accessed elements), and hits in B2 increase the size of T2 (proportion of cache dedicated to elements accessed at least twice). Evictions from T1 and T2 get added to B1 and B2, respectively.

3.3.2   SET DUELING

An issue with hybrid policies, such as ARC, is the large overhead of maintaining additional tag directories. Qureshi et al. [2006] introduce *Set Dueling*, which is an accurate mechanism for sampling the behavior of different policies at low hardware cost. Set Dueling builds on the observation that a few randomly chosen sets can accurately represent the behavior of different replacement policies on the entire cache. Qureshi et al. [2006] mathematically show that for caches with 1–4 MB (1024–4096 sets), 64 sets are enough to capture the behavior of the entire cache. We now discuss Set Dueling in more detail by discussing two representative policies.

Dynamic Insertion Policy (DIP)   The dynamic insertion policy (DIP) combines the benefit of recency-friendly policies and thrash-resistant policies by dynamically modulating the insertion positions of incoming lines [Qureshi et al., 2007]. In particular, DIP alternates between the recency-friendly LRU (Table 3.1) and the thrash-resistant Bimodal Insertion Policy (Table 3.6).

To choose between the two policies, DIP uses Set Dueling to dynamically track the hit rate of each policy. Figure 3.11 shows that DIP dedicates a few sets to LRU (sets 0, 5, 10, and 15 in Figure 3.11) and a few sets to BIP (sets 3, 6, 9, and 12 in Figure 3.11). These dedicated

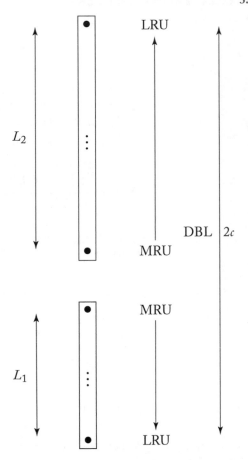

Figure 3.10: Adaptive replacement cache (ARC) maintains two tag directories; based on [Megiddo and Modha, 2003].

sets are called set dueling monitors (SDMs), while the remaining sets are called the *follower sets*. The policy selection (PSEL) saturating counter determines the winning policy by identifying the SDM that receives more cache hits. In particular, the PSEL is incremented when the SDM dedicated to LRU receives a hit, and it is decremented when the SDM dedicated to BIP receives a hit (a k-bit PSEL counter is initialized to 2^{k-1}). The winning policy is identified by the MSB of the PSEL. If the MSB of PSEL is 0, the follower sets use the LRU policy; otherwise the follower sets use BIP. Thus, Set Dueling does not require any separate storage structure other than the PSEL counter.

Dynamic Re-Reference Interval Policy (DRRIP)   DRRIP builds on DIP to add scan-resistance. In particular, DRRIP uses set dueling to create a hybrid of SRRIP, which is the

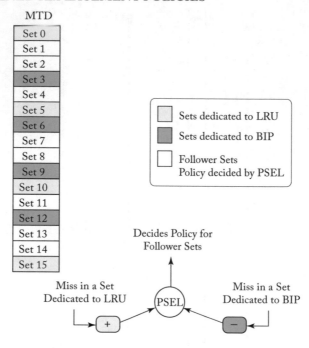

Figure 3.11: Set Dueling; based on [Qureshi et al., 2007].

scan-resistant version of LRU (Table 3.7), and BRRIP, which is the scan-resistant version of BIP (Table 3.8).

As we will see in Chapter 4, the insight behind Set Dueling has had a large impact on many subsequent Fine-Grained policies, which use the notion of set sampling to efficiently track metadata to determine fine-grained caching behavior.

CHAPTER 4

Fine-Grained Replacement Policies

Fine-Grained policies differentiate cache lines at the time of insertion, and this differentiation is typically based on eviction information from previous lifetimes of *similar* cache lines. For example, if a Fine-Grained policy learns that a line was evicted without being reused in its previous lifetimes, then it can insert the line into the cache with low priority. By contrast, a Coarse-Grained policy, such as LRU, will evict a line only after it has migrated from the MRU position to the LRU position, so it forces the line to reside in the cache for a long period of time—consuming precious cache space—just to determine that the line should be evicted. Thus, by learning from the behavior of previous cache lines, Fine-Grained policies can make more effective use of the cache.

We divide Fine-Grained policies into three broad categories based on the metric they use for predicting insertion priorities. The first category (Section 4.1) consists of solutions that predict expected reuse intervals for incoming lines. The second category (Section 4.2) consists of solutions that predict just a binary caching decision (cache-friendly vs. cache-averse). The third category, which is much smaller than the other two, includes policies [Beckmann and Sanchez, 2017, Kharbutli and Solihin, 2005] that introduces novel prediction metrics.

Fine-Grained solutions have several other design dimensions. First, since it can be cumbersome to remember the caching behavior of individual lines across multiple cache lifetimes, these policies learn caching behaviors for *groups of lines*. For example, many solutions group lines based on the address (PC) of the instruction that loaded the line, because lines that are loaded by the same PC tend to have similar caching behavior. Recent solutions look at more sophisticated ways to group cache lines [Jiménez and Teran, 2017, Teran et al., 2016]. A second design dimension is the amount of history that is used for learning cache behavior.

Fine-Grained replacement policies have roots in two seemingly different contexts. One line of work uses prediction to identify *dead blocks*—blocks that will not be used before being evicted—that could be re-purposed for other uses. For example, one of the earliest motivations for identifying dead blocks was to use them as prefetch buffers [Hu et al., 2002, Lai et al., 2001]. Another motivation was to turn off cache lines that are dead [Abella et al., 2005, Kaxiras et al., 2001]. The second line of work generalizes hybrid re-reference interval policies [Jaleel et al., 2010b] so that they are more learning based. Despite their different origins, these two lines of research have converged to conceptually similar solutions.

We now discuss the different classes of Fine-Grained policies.

4.1 REUSE DISTANCE PREDICTION POLICIES

Policies based on reuse distance estimate the *reuse distance* of blocks, where reuse distance can be defined in terms of the number of accesses or cycles between consecutive references to a block. The perfect prediction of reuse distances would be sufficient to emulate Belady's optimal solution, but it is difficult to precisely predict reuse distances due to the high variation in reuse distance values. Therefore, realistic solutions estimate reuse distance distributions or other aggregate reuse distance statistics.

4.1.1 EXPIRATION-BASED DEAD BLOCK PREDICTORS

Many dead block predictors use past evictions to estimate average reuse distance statistics, and they evict lines that are not reused within their expected reuse distances [Abella et al., 2005, Faldu and Grot, 2017, Hu et al., 2002, Kharbutli and Solihin, 2005, Liu et al., 2008, Takagi and Hiraki, 2004].

Hu et al. learn the *live time* of cache blocks, where the live time is defined as the number of cycles a block remains live in the cache [Hu et al., 2002]. When a block is inserted, its lifetime is predicted to be similar to its last lifetime. If the block has stayed in the cache for twice its lifetime without receiving a cache hit, the block is declared to be dead and is evicted from the cache.

Abella et al., use a similar dead block prediction strategy to turn off L2 cache lines, but instead of using the number of cycles, they define reuse distance in terms of the number of cache accesses between consecutive references [Abella et al., 2005].

Kharbutli and Solihin [2005] use counters to track each cache line's Access Interval, where a line's Access Interval is defined to be the number of other cache lines that were accessed between subsequent accesses to the line. Furthermore, they predict that a line is dead if its access interval exceeds a certain threshold. The threshold is predicted by the access interval predictor (AIP), which tracks the access intervals of all past evictions and remembers the maximum of all such intervals in a PC-based table.

Faldu and Grot [2017] present the Leeway policy, which uses the notion of *live distance*— the largest observed stack distance in a cache line's lifetime—to identify dead blocks. A cache block's live distance is learned from the block's previous generations, and when a block exceeds its live distance, it is predicted to be dead. The live distances from past lifetimes are remembered using a per-PC live distance predictor table (LDPT). The LDPT predicts live distances for incoming blocks, and any block that has moved past its predicted live distance in an LRU stack is predicted dead. To avoid the high variability in live distances across lifetimes and across blocks accessed by the same PC, Leeway deploys two update policies that control the rate at which live distance values in the LDPT are adjusted based on workload characteristics. The first policy is aggressive and favors bypassing, and the second policy is conservative and favors caching. Leeway uses Set Dueling [Qureshi et al., 2007] to choose between the two policies.

4.1.2 REUSE DISTANCE ORDERING

Keramidas et al. [2007] use reuse distance predictions to instead evict the line that is expected to be reused furthest in the future. Their policy learns a reuse distance for each load instruction (PC) and for each incoming line, it computes an estimated time of access (ETA), which is the sum of the current time and the expected reuse interval. It then orders lines based on their ETA and evicts the line with the largest ETA value.

To guard against cases where an ETA prediction is unavailable, this scheme also tracks the line's *decay*, which is an estimate of the amount of time that a line has remained unaccessed in the cache, and it evicts a line if its decay interval is larger than its ETA.

More concretely, Figure 4.1 shows that the replacement policy has two candidates: (1) the line with the largest ETA (the ETA line), and (2) the line with the largest Decay time (the LRU line). The line with the largest of the two values is selected for eviction. Thus, the policy relies on ETA when available and reverts to LRU otherwise.

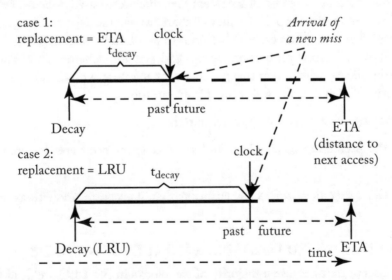

Figure 4.1: The ETA policy chooses between the line with the longest ETA or the one with the largest decay; based on [Keramidas et al., 2007].

4.2 CLASSIFICATION-BASED POLICIES

Classification-based policies learn a binary classification of incoming lines: is a cache access likely to result in a future hit or not? Cache-friendly lines—lines that are expected to result in cache hits—are inserted into the cache with a higher priority so that they have ample opportunity to receive a cache hit, and cache-averse lines—lines that are not expected to result in cache hits—are inserted with a low priority so that they can be quickly evicted without wasting cache resources.

Classification-based policies have several advantages over other classes of replacement policies. Compared to Hybrid replacement policies, which make a uniform decision for all lines in a given time period, Classification-based policies can insert some lines with high-priority and others with low-priority. Compared to Reuse Distance-based policies, where the target is a numerical reuse distance prediction, Classification-based policies solve a simpler binary prediction problem.

As we will see, Classification-based Policies have their origins in two different bodies of literature. While Sampling Based Dead Block Predictor (SDBP) [Khan et al., 2010] (Section 4.2.1) builds on the dead block prediction literature, Signature Based Hit Prediction (SHiP) [Wu et al., 2011a] (Section 4.2.2) has roots in the cache replacement policy literature. Interestingly, both solutions arrived at conceptually similar ideas and, in hindsight, the two papers have collectively unified the two areas.

We will now discuss these and other recent Classification-based policies. While there are significant differences among the policies, they all share two characteristics. First, they all include a binary predictor that learns past caching behavior to guide the insertion priority of individual lines. Second, they all borrow promotion, aging, and eviction schemes from state-of-the-art Coarse-Grained policies, which help them account for inaccurate predictions. As we describe these policies, we will consider the following design questions.

- Which caching solution is the policy learning?

- What is the prediction mechanism, and at what granularity are the predictions being made?

- What is the aging mechanism for ensuring that inaccurate predictions are eventually evicted?

4.2.1   SAMPLING BASED DEAD BLOCK PREDICTION (SDBP)

Many studies observe that because a majority of the blocks in the LLC are dead (they are not reused again before they are evicted), dead block prediction can be used to guide cache replacement and early bypass of dead blocks [Khan et al., 2010, Lai and Falsafi, 2000]. Lai and Falsafi [2000] introduce the idea of using dead block predictors to prefetch data into dead blocks in the L1. Their reftrace predictor predicts that if a trace of instruction addresses leads to the last access for one block, then the same trace will also lead to the last access for other blocks. To reduce the cost of maintaining an instruction trace for all cache blocks, Khan et al. [2010] introduce the sampling based dead block predictor (SDBP), which samples the caching behavior of program counters (PCs) to determine whether an incoming block is likely to be dead. Future cache accesses from PCs that are known to insert dead blocks are bypassed so that they do not pollute the cache. Accesses from PCs that do not insert dead blocks are inserted into the cache using some baseline policy, namely, a random or LRU replacement policy.

Notably, SDBP learns from a decoupled sampler that is populated using a small fraction of the cache accesses (see Figure 4.2). If a block is evicted from the sampler without being reused, the corresponding PC is trained negatively; otherwise, the predictor is trained positively. The decoupled sampler has several advantages. First, the predictor is trained using only a small sample of all cache accesses, which leads to a power- and space-efficient predictor design and manageable metadata in the sampler (PCs are maintained for each sampler entry). Second, the replacement policy of the sampler does not have to match the replacement policy of the cache. Khan et al. [2010] use the LRU policy for the sampler, and they use random replacement for the main cache. Finally, the associativity of the sampler is independent of the associativity of the cache, which allows for a cheaper sampler design. Khan et al. use a 12-way sampler for a 16-way cache. Table 4.1 summarizes SDBP's key operations.

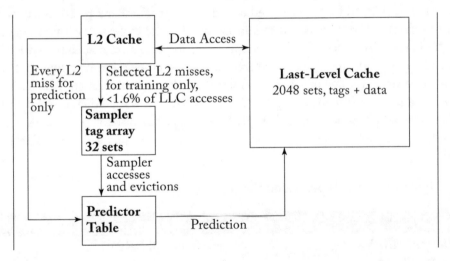

Figure 4.2: SDBP uses a decoupled sampler to train a predictor; based on [Khan et al., 2010].

Table 4.1: Sampling based dead block prediction (SDBP)

Insertion	Promotion	Aging	Victim Selection
dead_bit = *prediction*	dead_bit = *prediction*	Increment LRU	Line with dead_bit
MRU position	MRU position	counter by 1	or LRU position

Thus, to answer the questions that we outlined at the beginning of this section: (1) SDBP learns the caching decisions of an LRU-based sampler; (2) it predicts dead blocks (cache-averse vs. cache-friendly) using a skewed predictor design, making these predictions at the granularity of PCs; and (3) SDBP bypasses all incoming blocks that are predicted to be cache-averse, with the remaining lines being managed using the baseline replacement policy, so false positives (cache-averse blocks that are predicted to be cache-friendly) are aged out using the baseline replacement policy, and false negatives (cache-friendly blocks that are predicted to be cache-averse) do not get any opportunity to see reuse.

4.2.2   SIGNATURE BASED HIT PREDICTION (SHIP)

Like SDBP, SHiP [Wu et al., 2011a] learns the eviction behavior of the underlying replacement policy,[1] but the main insight behind SHiP is that reuse behavior is more strongly correlated with the PC that *inserted* the line into the cache rather than with the PC that last accessed the line. Thus, on a cache eviction, SHiP's predictor is trained with the PC that first inserted the line on a cache miss, and the predictor is only consulted on cache misses (on hits, lines are promoted to the highest priority without consulting the predictor).

More specifically, SHiP trains a predictor that learns whether a given signature has near or distant re-reference intervals. A few sets are sampled in the cache to maintain signatures and train the predictor. On a cache hit in a sampled set, the signature associated with the line is trained positively to indicate a near re-reference, and on an eviction of a line that was never reused, the signature is trained negatively to indicate a distant re-reference. When a new line is inserted, the signature of the incoming line is used to consult the predictor and determine the re-reference interval of the incoming line (prediction is performed for all accesses, not just the sampled sets). Once inserted into the cache, lines are managed using a simple RRIP policy (see Table 4.2).

Table 4.2: Signature based hit prediction (SHiP)

Insertion	Promotion	Aging	Victim Selection
If (*prediction*) RRPV = 2 else RRPV = 3	RRPV = 0	Increment all RRPV's (if no line with RRPV = 3)	RRPV = 3

The choice of signature is critical to SHiP's effectiveness. Jaleel et al. [2010b] evaluate a program counter signature (PC), a memory region signature, and an instruction sequence history signature, and they find that the PC signature performs best.

SHiP builds on DRRIP [Jaleel et al., 2010b] to create a Fine-Grained policy. Whereas DRRIP makes uniform re-reference predictions for all cache lines in an epoch, SHiP makes finer-grained predictions: it categorizes incoming lines into different groups by associating each reference with a unique *signature*. Cache lines with the same signature are assumed to have sim-

---

[1]A minor difference is that SDBP learns the LRU solution, whereas SHiP learns from SRRIP.

ilar re-reference behavior, but cache lines with different signatures are allowed to have different re-reference behavior within the same epoch.

We now answer the questions mentioned at the beginning of this sections. (1) Initially, SHiP learns from SRRIP, but once SHiP's predictor has been trained, further training updates come from SHiP's own reuse and eviction behavior. (2) SHiP uses a PC-based predictor, where the PC associated with a line is the one that inserted the line on a cache miss, where each PC is associated with a saturating counter. (3) SHiP relies on the RRIP policy to age all lines.

4.2.3   HAWKEYE

To avoid the pathologies of heuristic-based solutions, such as LRU, Hawkeye [Jain and Lin, 2016] builds off of Belady's MIN solution [Belady, 1966],[2] which is intriguing for two reasons. First, Belady's MIN is optimal for any sequence of references, so a MIN-based solution is likely to work for any access pattern. Second, Belady's MIN algorithm is an impractical algorithm because it replaces the line that will be reused furthest in the future; hence, it relies on knowledge of the future.

The key insight behind Hawkeye is that while it is impossible to look into the future, it *is* possible to apply Belady's MIN algorithm to the memory references of the past. Moreover, if a program's past behavior is a good predictor of its future behavior, then by learning the optimal solution for past, Hawkeye can train a predictor that should perform well for future accesses.

To understand how much history is needed to simulate MIN for past events, Jain and Lin [2016] study the performance of MIN by limiting its window into the future. Figure 4.3 shows that while MIN needs a long window into the future (8× the size of the cache for SPECint 2006), it does not need an unbounded window. Thus, to apply MIN to past events, we would need a history of 8× the size of the cache.

Since maintaining an 8× history is infeasibly expensive, Hawkeye computes the optimal solution for just a few sampled sets, and it introduces the OPTgen algorithm that computes the same answer as Belady's MIN policy for these sampled sets. OPTgen determines the lines that would have been cached if the MIN policy had been used. The key insight behind OPTgen is that the optimal caching decision of a line can be accurately determined when the line is next reused. Thus, on every reuse, OPTgen answers the following question: would this line have been a cache hit or cache miss with MIN? This insight enables OPTgen to be reproduce Belady's solution in $O(n)$ complexity using a small amount of hardware budget and simple hardware operations.

Figure 4.4 shows the overall design of the Hawkeye replacement policy. OPTgen trains the Hawkeye predictor, a PC-based predictor which learns whether lines inserted by a PC tend to be cache-friendly or cache-averse. When OPTgen determines that a line would have been a hit with MIN, the PC corresponding to the line is trained positively, otherwise it is trained

---

[2]Technically, the Hawkeye policy builds off of a new linear-time algorithm that produces the same result as Belady's MIN policy.

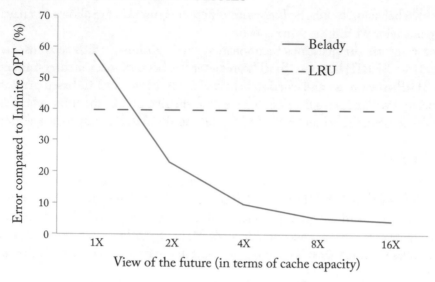

Figure 4.3: Belady's algorithm requires a long view of the future; used with permission [Jain and Lin, 2016].

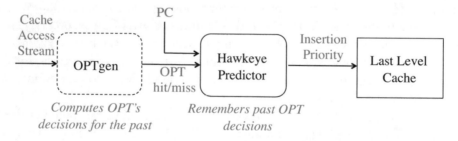

Figure 4.4: Overview of the Hawkeye cache; used with permission [Jain and Lin, 2016].

Table 4.3: Hawkeye

Insertion	Promotion	Aging	Victim Selection
If (*prediction*) RRPV = 0	(same as	Increment all RRPV's	RRPV = 7
else RRPV = 7	insertion)	(if no line with RRPV = 7)	

negatively. The predictor is consulted for every cache insertion and promotion[3] using the PC of the incoming line. Lines that are predicted to be cache-friendly are inserted with high priority, and lines that are predicted to be cache-averse are inserted with low priority (see Table 4.3).

[3]This differs from other Classification-based policies, where the predictor is only consulted on insertions.

To answer the questions at the beginning of the section: (1) Hawkeye learns from the optimal caching solution, instead of learning from LRU or SRRIP; (2) Hawkeye learns the optimal solution using a PC-based predictor; and (3) Hawkeye also relies on RRIP's aging mechanism to age lines that are inserted with high priority. To correct for inaccurate predictions, Hawkeye also trains the predictor negatively when a line that was predicted to be cache-averse is evicted without being reused.

Hawkeye makes an interesting conceptual contribution: it phrases cache replacement as a *supervised learning* problem,[4] which is surprising because unlike branch prediction, where the program execution eventually provides the correct outcomes of each branch, hardware caches do not provide such labeled data. By applying the optimal solution to past events, Hawkeye provides labeled data, which suggests that the field of cache replacement might benefit from the vast research in supervised learning.

4.2.4   PERCEPTRON-BASED PREDICTION

Predictive policies depend heavily on the accuracy of their predictors. SDBP, SHiP, and Hawkeye all use PC-based predictors that achieve accuracies of around 70–80%. Jiménez and Teran aim to improve predictor accuracy by using better features and better prediction models [Jiménez and Teran, 2017, Teran et al., 2016]. For example, the Perceptron Predictor [Teran et al., 2016] uses simple artificial neurons[5] [Rosenblatt, 1962] to augment the PC with richer features, such as (1) the history of PCs, (2) bits from the memory address, (3) a compressed representation of the data, and (4) the number of times a block has been accessed. Each feature is used to index a distinct table of saturating counters, which are then summed and compared against a threshold to generate a binary prediction. A small fraction of accesses are sampled to updated the perceptron predictor using the perceptron update rule: if the prediction is incorrect, or if the sum fails to exceed some magnitude, then the counters are decremented on an access and incremented on an eviction. Figure 4.5 contrasts the perceptron predictor (right) with prior PC-based predictors (left).

The Multiperspective Reuse Predictor [Jiménez and Teran, 2017] explores an extensive set of features that captures various properties of the program, producing a predictor that is informed from multiple perspectives. The features are parameterized with richer information about the LRU stack position of each training input, the bits of the PC with which each feature is hashed, and the length of the PC history. Together, these parameters create a large feature space that leads to higher prediction accuracy.

4.2.5   EVICTED ADDRESS FILTER (EAF)

The EAF policy [Seshadri et al., 2012] predicts the reuse behavior of each missed block individually, allowing for finer-grained differentiation than PCs. The key observation is that if a block

---

[4]In machine learning, supervised learning is the task of learning a function from a set of labeled input-output pairs.
[5]Perceptrons are the building blocks of neural networks.

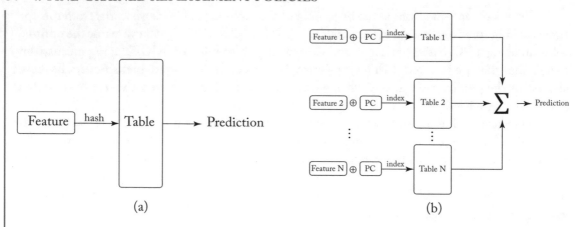

Figure 4.5: The perceptron predictor considers more features; based on [Teran et al., 2016].

with high reuse is prematurely evicted from the cache, then it will be accessed soon after eviction, while a block with low reuse will not be accessed for a long time after eviction. Therefore, the EAF policy uses a bloom filter [Bloom, 1970] (a probabilistic, space-efficient structure to determine the presence or absence of an element in a set) to track a small set of recently evicted addresses. On a cache miss, if the new line is present in the bloom filter—which is also called the Evicted Address Filter—then the line is predicted to be reuse-friendly and is inserted with high priority; otherwise, the new line is inserted according to the Bimodal Insertion policy (see Section 3.1.2). When the bloom filter is full, it is reset. The EAF is conceptually similar to a small victim cache that tracks lines that have been recently evicted from the main cache. Table 4.4 summarizes the operations of the EAF policy.

Table 4.4: The evicted address filter policy (EAF)

Insertion	Promotion	Aging	Victim Selection
MRU position (if in EAF), BIP (otherwise)	MRU position	Increment by 1	LRU position

Conceptually, the EAF policy extends the lifetime of cache lines beyond eviction, so that the lifetime of a line starts with its insertion, but it ends when the line is removed from the bloom filter. With an extended lifetime, it becomes feasible for EAF to observe reuse for lines with long reuse intervals, which leads to better scan-resistance and better thrash-resistance, and thus better performance.

The reuse detector (ReD) [crc, 2017, Albericio et al., 2013] proposes similar ideas as it bypasses any line that does not hit in the LLC or an *Address Reuse Table*, which tracks recent cache misses. As a result, ReD only inserts lines in the cache on their second reuse. To avoid bypassing all lines when they are first seen, ReD also uses a PC-based predictor to predict lines that are likely to be reused after their first access.

4.3   OTHER PREDICTION METRICS

Not all Fine-Grained policies predict reuse distances or binary labels, and it is certainly possible to capture past behavior using different prediction targets. For example, one component of Kharbutli and Solihin [2005] dead block predictor predicts the maximum number of times that a line is expected to be reused in the cache. As an example of this class of solutions, we now discuss in detail the EVA policy [Beckmann and Sanchez, 2017], which introduces a novel prediction target, called the EVA, and which is one of the few Fine-Grained solutions to use historical information to guide the aging process.

4.3.1   ECONOMIC VALUE ADDED (EVA)

Beckmann and Sanchez argue that it is optimal to replace the candidate with the longest expected time to reuse only when we possess perfect knowledge of the future [Belady, 1966], but this strategy is inadequate for practical solutions which face inherent uncertainty about the future [Beckmann and Sanchez, 2017]. Thus, practical solutions need to balance two competing objectives: (1) maximize the probability that a given line will hit in the cache, and (2) limit the duration for which the line consumes cache space. Solutions that are based solely on reuse distance account for only one side of the tradeoff.

To address this limitation, Beckmann and Sanchez [2017] propose a new metric called the *economic value added (EVA)*, which combines these two factors into a single metric. EVA is defined to be the number of hits the candidate is likely to yield compared to its average occupancy in the cache. Equation (4.1) shows EVA is computed for a given line. We see that there are two components to a line's EVA. First, the line is rewarded for its expected number of future hits (a line that has a higher reuse probability will have a higher EVA value). Second, the line is penalized for the cache space that it will consume. This penalty is computed by *charging* each candidate for the time it will spend in the cache at the rate of a single line's average hit rate (the cache's hit rate divided by its size), which is the long-term opportunity cost of consuming cache space. Thus, the EVA metric captures the cost-benefit tradeoff of caching a line by computing whether its odds of hitting are worth the cache space that it will consume.

$$EVA = Expected\_hits - (Cache\_hit\_rate/Cache\_size) \times Expected\_time. \qquad (4.1)$$

The EVA of candidates is inferred from their ages, and it is revised as the candidates age. For example, the left side of Figure 4.6 shows how the EVA changes with respect to a candidate's age for an application that iterates over a small and a big array (the reuse distance distribution for the same application is shown on the right side). At first, the EVA is high because there is

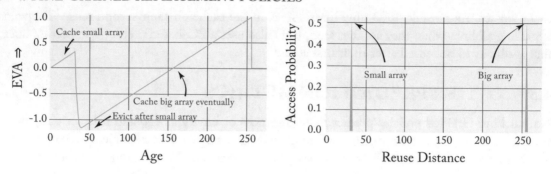

Figure 4.6: EVA learns about the candidates as they age; based on [Beckmann and Sanchez, 2017].

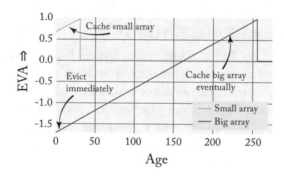

Figure 4.7: EVA with classification; based on [Beckmann and Sanchez, 2017].

a good chance that the access is from a small array, which means that the replacement policy can afford to gamble that candidates will hit quickly. But the EVA drops sharply when the age exceeds the length of the short array because the penalty of caching lines from the big array is much higher. At this point, the low EVA makes it very likely that lines from the big array are evicted. The penalty of caching lines from the big array decreases as the lines age and approach the reuse distance of the big array. This penalty is reflected in the gradual increase in the EVA from age 50 to age 250, at which point the EVA replacement policy protects lines from the big array even though they are old. Thus, in the face of uncertainty, the EVA replacement policy learns more about candidates as they age.

Of course, another way to reduce uncertainty is to classify lines into different categories. For example, Figure 4.7 shows the EVA with respect to age if we were to classify the small array and big array into distinct classes, and we see that the per-class EVA curves are much simpler. Theoretically, EVA can be extended to support classification, but the implementation complexity limits this extension to a few classes. In the next section, we will discuss replacement policies that rely on many fine-grained classes but learn much simpler metrics for each class. By contrast, EVA ranks ages at a fine granularity, but this restricts EVA to use fewer classes.

The EVA replacement policy computes the EVA curves by recording the age distribution of hits and evictions and by processing information about these evictions using a lightweight software runtime. To predict the EVA of a line, its age is used to index into an *eviction priority array*, which conceptually represents the EVA curves.

CHAPTER 5

Richer Considerations

Until now we have focused narrowly on the problem of cache replacement, both in terms of metrics—in which we have focused on cache misses—and in terms of context—in which we have focused on the cache as an isolated abstract entity. But, of course, cache misses do not translate directly to performance loss, and cache replacement policies do not operate in isolation. This chapter broadens our discussion in both dimensions, first exploring replacement policies that consider the variable cost of cache misses and then exploring policies that consider the impact of prefetchers, the impact of the cache architecture, and finally the impact of new memory technologies.

5.1 COST-AWARE CACHE REPLACEMENT

Most replacement policies seek to minimize overall cache miss rate, assuming that all cache misses are equally costly. In reality, different cache misses can have widely different impact on overall performance. For example, misses that are isolated (low memory-level parallelism) tend to be more expensive than misses that are clustered (high memory-level parallelism), because with greater parallelism there is more ability to hide the latency of a single cache miss. As another example, misses that are on the program's critical path are more important to program performance than those that are not. A smart replacement policy that considers these costs can preferentially cache high-cost misses (at the cost of a lower hit rate) to achieve better program performance.

The optimal cost-aware replacement policy (CSOPT) [Jeong and Dubois, 2006] minimizes the overall cost of all misses, rather than minimizing the number of misses. The basic idea is to follow Belady's MIN policy except when MIN would evict a block whose miss cost is greater than those of other blocks in the cache. On encountering such a block, CSOPT explores multiple sequences of replacement decisions until it can decide on the replacement sequence that minimizes total cost. A naive realization of CSOPT expands all possible replacement sequences in the form of a search tree (see Figure 5.1); the depth of the search tree is equal to the number of cache accesses, because a new level is added for every new cache access, and the width is equal to the cache size s, because the search considers up to s replacement choices on every cache eviction.

Note that a search tree is required because greedily caching a high-cost block at any node in the tree might not translate to the best replacement sequence in the long run. In particular, greedily caching a high-cost block might preclude the caching of another block that has an even

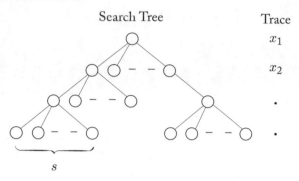

Figure 5.1: CSOPT uses a search tree; based on [Jeong and Dubois, 2006].

higher cost, or it might displace several low-cost blocks that together are more costly than the single high-cost block. Thus, the best replacement decision depends on the behavior of multiple competing lines in the future, resulting in a large search space of solutions.

Unfortunately, CSOPT's branch-and-bound approach is exponential in the number of cache accesses, whereas MIN is linear in the number of cache accesses. Jeong and Dubois [2006] propose heuristics to prune the search tree and to reduce the search space. For a few scientific workloads, these heuristics are shown to make the computation of CSOPT tractable—in the sense that an offline analysis can complete—but in general, they do not reduce the worst-case complexity.

Figure 5.2 shows the relative cost savings of CSOPT over MIN (y-axis) for the Barnes Hut benchmark from the SPLASH benchmark suite [Jeong and Dubois, 2006]. Here cache misses are assumed to have just two possible costs, either high cost or low cost. The x-axis represents different proportions of high-cost accesses, and the lines represent different cost ratios between high-cost and low-cost accesses. We see that the cost savings of CSOPT over MIN increases with higher cost ratios, and they are most pronounced when the high-cost accesses comprise 20–30% of the total number of accesses. This trend makes sense because the benefit of CSOPT over MIN grows when the difference is cost is high and when the percentage of low-cost misses is high, because in these cases, MIN is more likely to make the wrong choice.

Of course, like MIN, CSOPT is impractical since it requires knowledge of future accesses. Therefore, practical solutions for cost-aware cache replacement rely on intuitive heuristics to (1) identify high-cost cache accesses and (2) preferentially cache high-cost loads over low-cost loads.

5.1.1   MEMORY-LEVEL PARALLELISM (MLP)

Qureshi et al. [2006] were the first to propose an MLP-aware cache replacement policy. Their key insight is that isolated misses (low MLP) are more costly for performance than parallel misses (high MLP) because an isolated miss impedes all dependent instructions behind it and

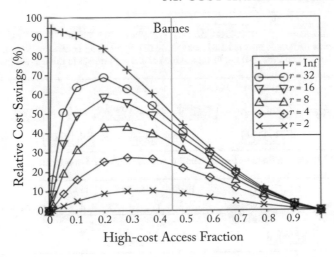

Figure 5.2: Cost savings over MIN with random cost assignments [Jeong and Dubois, 2006].

leads to a long-latency processor stall. By contrast, parallel misses are not as expensive because the processor can hide their latency by issuing multiple loads in parallel. Thus, an MLP-aware cache replacement policy can improve performance by reducing the number of performance-critical isolated misses.

Qureshi et al. illustrate this phenomenon with an example (see Figure 5.3), where $P1$, $P2$, $P3$, and $P4$ can be serviced in parallel, and where $S1$, $S2$, and $S3$ are isolated misses. The top part of Figure 5.3 shows that Belady's solution results in only 4 misses, but it also results in 4 stalls because the $S1$, $S2$, and $S3$ all miss and cause the processor to stall. By contrast, the MLP-aware policy in the bottom part of the figure never evicts $S1$, $S2$, and $S3$, so even though this policy results in six misses, it stalls the processor only twice, resulting in overall better performance.

MLP-Aware Linear Replacement Policy There are two components to Qureshi et al. [2006]'s MLP-aware Linear (LIN) policy. The first is an algorithm that predicts the MLP-cost of a future miss. The second is a replacement policy that takes into account both recency and MLP-based cost.

The MLP-based cost of a cache miss is defined to be the number of cycles that the miss spends waiting to be serviced, which can be easily tracked using Miss Status Handling Registers. For parallel misses, the cycle count is divided equally among all concurrent misses. Thus, a higher MLP-based cost implies that the line is likely to result in an isolated miss, so it is more desirable to keep it in the cache. Furthermore, Qureshi et al. note that the MLP-based cost repeats for consecutive misses, so the last MLP-based cost of a miss is a good indicator of its MLP-based cost for future misses.

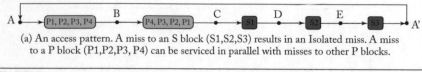

(a) An access pattern. A miss to an S block (S1,S2,S3) results in an Isolated miss. A miss to a P block (P1,P2,P3, P4) can be serviced in parallel with misses to other P blocks.

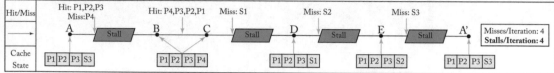

(b) Execution timeline for one iteration with Belady's OPT replacement.

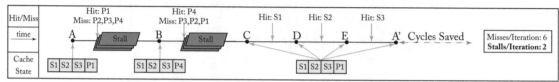

(c) Execution timeline for one iteration with MLP-aware replacement.

Figure 5.3: MLP-aware caching can outperform OPT; based on [Qureshi et al., 2006].

The LIN replacement policy linearly combines recency and MLP-based cost and evicts the line with the lowest aggregate cost. In particular, if $R(i)$ is the recency value of a block i (the highest value denotes the MRU position and lowest value denotes the LRU position), and $cost_q(i)$ is its quantized MLP-based cost, then the LIN policy will evict the line with the lowest value of $R(i) + \lambda * cost_q(i)$, where λ is the importance of the $cost_q$ in choosing the victim. A high value of λ indicates that LIN is likely to retain recent blocks with a high MLP-based cost, and a low value of λ indicates that LIN is likely to put more emphasis on recency. Qureshi et al. set λ to 4.

Finally, because LIN does not always outperform LRU, Qureshi et al. dynamically choose between LRU and LIN by using Set Dueling (see Section 3.3.2) to periodically choose between the LIN and LRU policies. Arunkumar and Wu [2014] provide an alternate solution by incorporating post-eviction reuse information into the cost-metric itself. In particular, their ReMAP policy defines the overall cost of a line as a linear combination of recency, DRAM latency, and post-eviction reuse distance, where the post-eviction reuse distance is computed using a bloom filter for evicted lines.

Locality-Aware Cost-Sensitive Cache Replacement Policy   Recent work builds on Qureshi et al.'s work by (1) defining new cost metrics, and (2) defining new replacement strategies. For example, the Locality-Aware Cost-Sensitive Cache Replacement Algorithm (LACS) [Kharbutli and Sheikh, 2014] estimates the cost of a cache block by counting the number of instructions issued during a block's LLC miss. Intuitively, this definition of cost reflects the processor's ability to hide miss penalty, which is similar to MLP. Cache blocks are classified as either low-cost or high-cost, depending on whether the number of issued instructions is above or below a threshold. For each block, LACS maintains a 2-bit cost value so that they can represent both high and low cost with two levels of confidence. Thus, instead of defining a numeric cost value, LACS uses a binary cost classification.

For its replacement policy, LACS attempts to reserve high-cost blocks in the cache, but only while their locality is still high (i.e., they have been accessed recently). In particular, on a cache miss, LACS chooses a low-cost block to evict, so that high-cost blocks remain in the cache. However, high-cost blocks are aged by decrementing their 2-bit cost value so that they relinquish the reservation if they've been in the cache for too long. Similarly, low-cost blocks are promoted by incrementing their 2-bit cost value so that they are retained in the cache if they show high temporal locality. Thus, LACS combines the benefits of both locality and cost-based replacement.

5.2   CRITICALITY-DRIVEN CACHE OPTIMIZATIONS

Criticality is a more general cost function than MLP: Srinivasan et al. [2001] define a critical load as any load that needs to complete early to prevent processor stalls, while a non-critical load is one that can tolerate long latency. Critical loads are identified using a variety of techniques [Fields et al., 2001, Srinivasan and Lebeck, 1998], and criticality-driven cache optimizations prioritize critical loads over non-critical loads. To highlight important advances in this area, we now discuss two criticality-driven optimizations.

5.2.1   CRITICAL CACHE

Using detailed limit studies, Srinivasan and Lebeck [1998] determine that load criticality can be determined by the characteristics of the chain of instructions dependent on the load. In particular, a load is classified as critical if it satisfies any of the following criteria: (1) the load feeds into a mispredicted branch, (2) the load feeds into another load that incurs an L1 cache miss, or (3) the number of independent instructions issued in an N cycle window following the load is below some threshold. Thus, this definition of criticality considers both the type of dependent instructions (e.g., mispredicted branch, L1 misses) and the number of instructions in its dependence chain.

To illustrate the importance of optimizing for critical loads, Srinivasan and Lebeck [1998] show that if all critical loads could be satisfied by the L1 cache, the result would be an average 40% improvement over a traditional memory hierarchy, whereas if an equal percentage of loads is randomly chosen to hit in the L1, the average improvement would be only 12%. Therefore, it may be possible to improve overall performance by decreasing the latency of these critical loads at the expense of increased latency for non-critical loads.

To optimize critical loads, Srinivasan and Lebeck [1998] use a *critical cache*, which serves as a victim cache for blocks that were touched by a critical load. For a fair comparison, the baseline also includes a victim cache, called the *locality cache*, which caches both critical and non-critical victims. Unfortunately, they find that the critical cache does not produce any gains over the locality cache because (1) the working set of critical loads is so large that the critical cache is unable to significantly reduce the critical load miss ratio, (2) the locality cache is able to provide competitive critical load miss ratios as a critical cache because of spatial locality between non-critical and critical loads, and (3) the critical cache's benefits are diminished by its overall higher miss rate. The overall result is that managing cache content based on locality outperforms criticality-based techniques.

Srinivasan and Lebeck [1998] conclude that it is very difficult to build memory hierarchies that violate locality to exploit criticality because there is a tension between increase in non-critical miss ratio and a decrease in critical miss ratio. However, they suggest that there is room for criticality-based techniques to supplement locality instead of replacing locality.

5.2.2   CRITICALITY-AWARE MULTI-LEVEL CACHE HIERARCHY

Nori et al. [2018] propose a criticality aware tiered cache hierarchy (CATCH) that accurately detects program criticality in hardware and uses a novel set of inter-cache prefetchers to ensure that critical loads are served at the latency of the L1 cache.

Nori et al. use a definition of criticality based on the program's data dependence graph (DDG), which was first introduced by Fields et al. [2001] in 2001. The key idea is to analyze the program's DDG to find the critical path, which is the path with the maximum weighted length, where weights are based on the latency of operations. All load instructions on the critical path are deemed critical. Figure 5.4 shows an example of a DDG, where each instruction in the DDG has three nodes: the D node denotes allocation into the core, the E node denotes the dispatch of the instruction to the execution nodes, and the C node denotes instruction commit. An E-E edge denotes a data dependence, C-C edges denote in-order commit, and D-D nodes denote inorder allocation into the core. The critical path in the graph is marked with dotted lines, and instructions 1, 2, 4, and 5 are found to be critical. Nori et al. propose a novel and

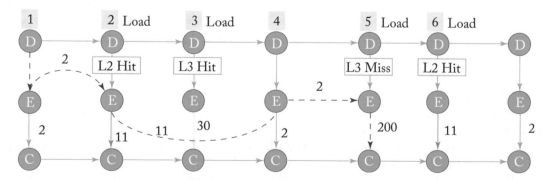

Figure 5.4: Critical path in a program's data dependence graph (DDG); based on [Nori et al., 2018].

fast incremental method to learn the critical path using an optimized representation of the data dependence graph in hardware, which takes just 3 KB of area. This method is used to enumerate a small set of critical load instructions (critical PCs).

To optimize critical loads, critical loads are prefetched from the L2 or the LLC into the L1 cache with the help of timeliness aware and criticality triggered (TACT) prefetchers. TACT utilizes the association between the address or data of load instructions in the vicinity of the target critical load to trigger the prefetches into L1. The implication of TACT prefetchers is that critical loads in the LLC can be served at the latency of the L1 cache, and the L2 cache is no longer necessary and can be eliminated altogether.

5.3   MULTI-CORE-AWARE CACHE MANAGEMENT

In multi-core systems, accesses from multiple cores compete for shared cache capacity. Poor management of the shared cache can degrade system performance and result in unfair allocation of resources because one ill-behaved application can degrade the performance of all other applications sharing the cache. For example, a workload with streaming memory accesses can evict useful data belonging to other recency-friendly applications.

There are two broad approaches for handling shared cache interference. The first approach partitions the cache among cores to avoid interference and to provide fairness guarantees. The second approach modifies the replacement policy to avoid pathological behavior. We now discuss both these approaches in more detail.

5.3.1   CACHE PARTITIONING

Partitioning avoids performance pathologies in shared caches and can provide strong isolation and fairness guarantees. Cache partitioning schemes have two main considerations. (1) How are partition sizes enforced in the cache? (2) How are partition sizes determined?

The most common mechanism to enforce cache partitions is to allocate dedicated ways to each application, such that any given application can only insert and evict from ways that are currently allocated to its partition. More advanced schemes avoid rigid partitions and instead modify the replacement policy to ensure that partitions are enforced in the average case [Sanchez and Kozyrakis, 2011, Xie and Loh, 2009].

Partition sizes can be determined by either the user, the OS, or the hardware. We now discuss two hardware-based schemes to determine partition sizes for shared last-level caches.

Utility-Based Cache Partitioning (UCP)   UCP is built on the insight that LRU does not work well for shared caches because it tends to allocate the most cache capacity to the application that issues the most memory requests rather than to the application that benefits the most from the cache [Qureshi and Patt, 2006]. To address this issue, Qureshi and Patt dynamically partition the cache among cores, and they propose a lightweight runtime mechanism to estimate the partition size for each core. The key idea behind the partitioning scheme is to allocate a larger cache partition to the application that is more likely to see an improved hit rate with the extra cache space. For example, in Figure 5.5, LRU gives equake 7 cache ways even though it does not see any hit rate improvement beyond 3 ways.

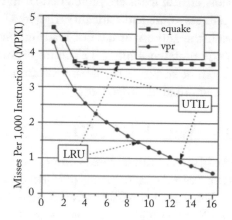

Figure 5.5: UCP allocates partition sizes based on utility; based on [Qureshi and Patt, 2006].

To find the hit rate curve for each competing application, UCP leverages the stack property of the LRU policy, which says that an access that hits in a LRU managed cache containing n ways is guaranteed to also hit if the cache had more than n ways (assuming that the number of sets remains constant). In particular, UCP deploys *sampled* auxiliary tag directories, called Utility Monitors or UMONs, to monitor the reuse behavior for each application, assuming that it had the entire cache to itself; for each application, it counts the number of cache hits at each recency position. The number of hits at each recency position determines the marginal utility of giving the application one extra way. For example, if for a cache with 2 ways, 25 of 100 accesses miss the cache, 70 hit in the MRU position, and only 5 hit in the LRU position, then reducing the cache size from two ways to one way will increase the miss count from 25–30. However, reducing the cache size from one way to zero ways will dramatically increase the miss count from 30–100.

UCP's partitioning algorithm uses the information collected by the UMONs to determine the number of ways to allocate to each core, and it uses way partitioning to enforce these partitions.

ASM-Cache   Since hit rates do not correspond directly to improved performance, ASM-Cache partitions shared caches with the goal of minimizing application slowdown [Subramanian et al., 2015]. The key idea is to allocate more cache ways to applications whose slowdowns reduce the most from additional cache space.

An application's slowdown is defined to be the ratio of its execution time when it is run with other applications (shared execution time) and its execution time had it been run alone on the same system (alone execution time). Since the shared execution time can be measured by sampling the current execution, the key challenge for ASM-Cache is to estimate the application's alone execution time without actually running it alone.

To estimate the application's alone time, ASM-Cache leverages the insight that the performance of a memory-bound application is strongly correlated with the rate at which its shared cache accesses are serviced. Therefore, ASM-cache estimates the shared cache service rate by (1) minimizing interference at the main memory to estimate average miss service time in the absence of interference and (2) measuring the number of shared cache misses in the absence of interference. The former is accomplished by temporarily giving the application high priority at the memory controller, and the latter is accomplished by using sampled auxiliary tags that only service one application. The aggregate miss count in the sampled tags is used along with the average miss service time to estimate the time it would have taken to serve the application's requests had it been run alone.

Using this scheme, ASM-Cache is able to estimate the slowdown of each application with different number of cache ways. The actual number of ways allocated to each application uses a scheme similar to UCP.

5.3.2 SHARED-CACHE-AWARE CACHE REPLACEMENT

Replacement policies discussed in Chapter 3 and Chapter 4 can be directly applied to shared caches. In general, early Coarse-Grained policies, such as LRU, that were susceptible to thrashing perform poorly in the shared cache setting, but more recent Fine-Grained policies tend to scale well without significant changes. The success of Fine-Grained replacement policies on shared caches can perhaps be explained by their ability to distinguish access patterns for small groups of lines, which allows them to naturally distinguish the behavior of different applications.

We now discuss two replacement schemes that were proposed to allow Coarse-Grained policies to better adapt to shared caches, and we also discuss a domain-specific replacement policy for task-flow programs.

Thread-Aware DIP The dynamic insertion policy (DIP) [Qureshi et al., 2007] discussed in Section 3.3 uses Set Dueling to modulate between the recency-friendly LRU policy and the thrash-resistant BIP policy. However, DIP leaves room for improvement in shared caches as it does not distinguish between the caching behavior of individual applications. So if DIP chooses LRU, all applications sharing the cache will use the LRU policy, which is not desirable when some applications benefit from the cache and others don't.

Jaleel et al. [2008] propose thread-aware extensions to DIP that are designed to work well for shared caches. The key idea is to make a choice for each application individually. However, this idea is complicated by the fact that for n applications sharing the cache, sampling the behavior of 2^n combinations is prohibitively expensive. To mitigate this overhead, they propose two approaches. The first approach, called TADIP-Isolated (TADIP-I), learns the insertion policy for each application independently, assuming that all other applications use the LRU policy. The second approach, called TADIP-Feedback (TADIP-F), accounts for interaction among applications by learning the insertion policy for each application, assuming that all other applications use the insertion policy that currently performs the best for that application.

Promotion/Insertion Pseudo-Partitioning of Multi-Core Shared Caches (PIPP) Xie and Loh [2009] build on Utility-Based Cache Partitioning [Qureshi and Patt, 2006], but instead of strictly enforcing UCP partitions, they design insertion and promotion policies that enforce the partitions loosely. The main insight behind their PIPP policy is that strict partitions result in under-utilization of cache resources because a core might not use its entire partition. For example, if the cache is way-partitioned, and if $core\_i$ does not access a given set, the ways allocated to $core\_i$ in that set will go to waste. PIPP allows other applications to steal these unused ways.

In particular, PIPP inserts each line with a priority that is determined by its partition allocation. Lines from cores that have been allocated large partitions are inserted with high priority (proportional to the size of the partition), and lines from cores that have been allocated small partitions are inserted with low priority. On a cache hit, PIPP's promotion policy promotes the line by a single priority position with a probability of p_prom, and the priority is unchanged with a probability of $1 - p\_prom$. On eviction, the line with the lowest priority is evicted.

RADAR Our discussion so far has focused on multiple programs sharing the last-level cache. Manivannan et al. [2016] instead look at the problem of last-level cache replacement for task-parallel programs running on a multi-core system. Their policy, called RADAR, combines static and dynamic program information to predict dead blocks for task-parallel programs. In particular, RADAR builds on task-flow programming models, such as OpenMP, where programmer annotations explicitly specify (1) dependences between tasks and (2) address regions that will be accessed by each task. The runtime system uses this information in conjunction with dynamic program behavior to predict regions that are likely to be dead. Blocks that belong to dead regions are demoted and preferentially evicted from the cache.

More concretely, RADAR has three variants that combine information from the programming model and the architecture in different ways. First, the Look-ahead scheme uses the task data-flow graph to peek into the window of tasks that are going to be executed soon, and it uses this information to identify regions that are likely to be accessed in the future and regions that are likely to be dead. Second, the Look-back scheme tracks per-region access history to predict when the next region access is likely to occur. Finally, the combined scheme exploits knowledge of future region accesses and past region accesses to make more accurate predictions.

5.4 PREFETCH-AWARE CACHE REPLACEMENT

In addition to caches, modern processors use prefetchers to hide the long latency of accessing DRAM, and it is essential that these mechanisms work well together. Prefetched data typically forms a significant portion of the cache, so, interaction with the prefetcher is an important consideration for cache replacement policies.

There are two primary goals in designing a prefetch-aware cache replacement policy. First, the replacement policy should avoid *cache pollution* caused by inaccurate prefetches. Second, the replacement policy should preferentially discard lines that can be prefetched over those that are difficult to prefetch [Jain and Lin, 2018, Srinath et al., 2007, Wu et al., 2011b].

In this section, we first summarize the vast majority of work in cache replacement which focuses on the first design goal of eliminating useless prefetches. We then show that Belady's MIN algorithm, which is provably optimal in the absence of prefetches, is incomplete in the presence of a prefetcher because it ignores the second design goal of deprioritizing prefetchable lines. Finally, we summarize recent work that addresses these limitations of MIN by simultaneously considering both of the above design goals.

5.4.1 CACHE POLLUTION

Most prefetch-aware cache replacement policies focus on reducing cache pollution by identifying and evicting inaccurate prefetches. Such solutions can be divided into two broad categories.

The first category takes feedback from the prefetcher to identify potentially inaccurate prefetch requests; such policies typically require explicitly co-designed prefetchers or modifications to existing prefetchers. For example, Ishii et al. use the internal state of the AMPM [Ishii et al., 2011] prefetcher to inform the insertion priority of prefetched blocks [Ishii et al., 2012]. As another example, Kill-the-PC (KPC) [Kim et al., 2017] is a cooperative prefetching and caching scheme that co-designs the prefetcher to provide feedback on confidence and estimated time to reuse. This information is then used to determine whether the prefetch is inserted into the L2 or the L3, and this information is used to determine the inserted line's RRPV insertion position.

The second category works independently of the prefetcher and monitors cache behavior to adapt replacement decisions; such policies can work with any prefetcher but may lack precise prefetcher-specific information. For example, Feedback Directed Prefetching (FDP) [Srinath et al., 2007] and Informed Caching policies for Prefetched blocks (ICP) [Seshadri et al., 2015] introduce methods to dynamically estimate prefetch accuracy, and they insert inaccurate prefetches at positions with low priority. In particular, FDP estimates accuracy at a coarse granularity by counting the ratio of useful prefetches and total prefetches within a time epoch. ICP augments FDP by accounting for recently evicted prefetched blocks that would have been deemed accurate if they had been cached for some additional amount of time. Prefetch-Aware Cache Management (PACMan) [Wu et al., 2011b] instead identifies for each time epoch the best insertion and promotion policies for prefetch requests. As shown in Figure 5.6, PACMan defines three variants of RRIP (PACMan-M, PACMan-H, and PACMan-HM), and it uses set dueling to find the best insertion policy for a given epoch.

	Baseline DRRIP	PACMan-M on DRRIP		PACMan-H on DRRIP		PACMan-HM on DRRIP	
SRRIP	**All**	**Demand**	**Prefetch**	**Demand**	**Prefetch**	**Demand**	**Prefetch**
Insertion	2	2	3	2	2	2	3
Re-Reference	0	0	0	0	No Update	0	No Update
BRRIP	**All**	**Demand**	**Prefetch**	**Demand**	**Prefetch**	**Demand**	**Prefetch**
Insertion	Mostly 3	Mostly 3	Mostly 3	Mostly 3	Mostly 3	Mostly 3	Mostly 3
Re-Reference	0	0	0	0	No Update	0	No Update

Figure 5.6: PACMan's RRIP policies; based on [Wu et al., 2011b].

Upon closer inspection of PACMan's three constituent policies, we see that while PACMan-M focuses on avoiding cache pollution by inserting prefetches in the LRU position, and PACMan-H *deprioritizes* prefetchable lines by not promoting prefetch requests on cache hits. Thus, PACMan-H is the first instance of a replacement policy that attempts to retain hard-to-prefetch lines (our second goal), but as we show in the next section, a much richer space of solutions exist when distinguishing between prefetchable and hard-to-prefetch lines.

5.4.2   DEPRIORITIZING PREFETCHABLE LINES

Belady's MIN is incomplete in the presence of prefetches because it does not distinguish between prefetchable and hard-to-prefetch lines.[1] In particular, MIN is equally inclined to cache lines whose next reuse is due to a prefetch request (prefetchable lines) and lines whose next reuse is due to a demand request (hard-to-prefetch lines). For example, in Figure 5.7, MIN might cache X at both $t = 0$ and $t = 1$, even though the demand request at $t = 2$ can be serviced by only caching X at $t = 1$. As a result, MIN minimizes the total number of cache misses, including those for prefetched lines (such as the request to X at $t = 1$), but it does not minimize the number of demand misses [Jain and Lin, 2018].

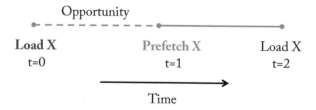

Figure 5.7: Opportunity to improve upon MIN; used with permission [Jain and Lin, 2018].

[1]MIN correctly handles cache pollution, as inaccurate prefetches are always reused furthest in the future.

Demand-MIN   To address this limitation, Jain and Lin propose a variant of Belady's MIN, called Demand-MIN, that minimizes demand misses in the presence of prefetches. Unlike MIN, which evicts the line that is reused furthest in the future, Demand-MIN evicts the line that is *prefetched* furthest in the future. More precisely, Demand-MIN states:

> *Evict the line that will be prefetched furthest in the future, and if no such line exists, evict the line that will see a demand request furthest in the future.*

Thus, by preferentially evicting lines that can be prefetched in the future, Demand-MIN accommodates lines that cannot be prefetched. For example, in Figure 5.7, Demand-MIN recognizes that because line X will be prefetched at time $t = 1$, line X can be evicted at $t = 0$, thereby freeing up cache space in the time interval between $t = 0$ and $t = 1$, which can be utilized to cache other demand loads. The reduction in demand miss rate can be significant: on a mix of SPEC 2006 benchmarks running on 4 cores, LRU yields an average MPKI of 29.8, MIN an average of 21.7, and Demand-MIN an average of 16.9.

Unfortunately, Demand-MIN's increase in demand hit rate comes at the expense of a larger number of *prefetch misses*,[2] which results in extra prefetch traffic. Thus, MIN and Demand-MIN define the extreme points of a design space, with MIN minimizing overall traffic on one extreme and Demand-MIN minimizing demand misses on the other.

Design Space   Figure 5.8 shows the tradeoff between demand hit rate (x-axis) and overall traffic (y-axis) for several SPEC benchmarks [Jain and Lin, 2018]. We see that different benchmarks will prefer different points in this design space. Benchmarks such as astar (blue) and sphinx (orange) have lines that are close to horizontal, so they can enjoy the increase in demand hit rate that Demand-MIN provides while incurring little increase in memory traffic. By contrast, benchmarks such as tonto (light blue) and calculix (purple) have vertical lines, so Demand-MIN increases traffic but provides no improvement in demand hit rate. Finally, the remaining benchmarks (bwaves and cactus) present less obvious tradeoffs.

To navigate this design space, Flex-MIN picks a point between MIN and Demand-MIN, such that the chosen point has a good tradeoff between demand hit rate and traffic. In particular, Flex-MIN is built on the notion of a *protected line*, which is a cache line that would be evicted by Demand-MIN but not by Flex-MIN because it would generate traffic without providing a significant improvement in hit rate. Thus, Flex-MIN is defined as follows:

> *Evict the line that will be prefetched furthest in the future and is not protected. If no such line exists, default to MIN.*

Jain and Lin [2018] define a simple heuristic to identify protected lines. Of course, unlike MIN and Demand-MIN, Flex-MIN is not optimal in any theoretical sense since it's built on a heuristic.

---

[2]Prefetch misses are prefetch requests that miss in the cache.

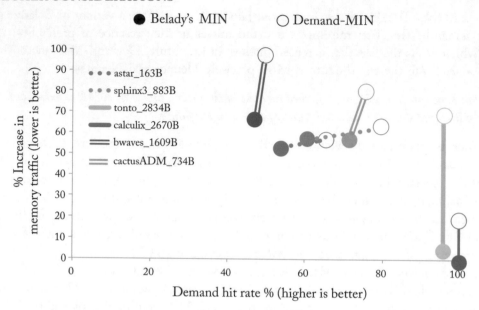

Figure 5.8: With prefetching, replacement policies face a tradeoff between demand hit rate and prefetcher traffic; used with permission [Jain and Lin, 2018].

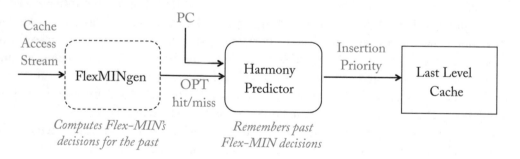

Figure 5.9: Harmony cache replacement policy; used with permission [Jain and Lin, 2018].

Harmony   Harmony is a practical replacement policy that explores the rich design space between MIN and Demand-MIN by learning from Flex-MIN. Harmony's overall structure (see Figure 5.9) is similar to Hawkeye's [Jain and Lin, 2016] (see Section 4.2), but the main difference is that Harmony replaces OPTgen with FlexMINgen, where FlexMINgen emulates Flex-MIN's solution. Like Hawkeye, Harmony's predictor is also PC-based, except Harmony has two predictors, one for demand requests and one for prefetch requests.

5.5   CACHE ARCHITECTURE-AWARE CACHE REPLACEMENT

So far, we have assumed that the priority ordering inferred by the cache replacement policy is independent of the cache architecture. However, changes in cache architecture can have implications for cache replacement. We now discuss two such changes to the cache architecture.

5.5.1   INCLUSION-AWARE CACHE REPLACEMENT

Inclusive caches require that the contents of all the smaller caches in a multi-level cache hierarchy be a subset of the LLC. Inclusion greatly simplifies the cache coherence protocol [Baer and Wang, 1988], but it limits the effective capacity of the cache hierarchy to be the size of the LLC (as opposed to the sum of all cache levels in an exclusive cache hierarchy).

Jaleel et al. [2010a] show that in an inclusive cache hierarchy, the LLC replacement policy has a significant impact on the smaller caches. In particular, when a line is evicted from the LLC, it is invalidated in the small caches to enforce inclusion, and these *inclusion victims* deteriorate the hit rate of the smaller caches because they can potentially evict lines with high temporal locality. In fact, Jaleel et al. show that the first order benefit of non-inclusion is the elimination of inclusion victims and not the extra cache capacity.

To avoid this pathology in inclusive cache hierarchies, Jaleel et al. propose three cache replacement policies to preserve hot lines in the smaller caches and to extend the lifetime of such caches in the LLC. The first policy, called temporal locality hints (TLH), conveys the temporal locality of "hot" lines in the core caches by sending hints to the LLC. These hints are used to update the replacement state in the LLC so that the LLC is less likely to choose a victim that will force an inclusion victim. The second policy, called early core invalidation (ECI) derives the temporal locality of the line in smaller caches while the line still has high priority in the LLC. The main idea is to choose a line early, invalidate it in the smaller caches while retaining it in the LLC; subsequent requests to the LLC for that line indicate the temporal locality of the line for smaller caches. Finally, the third policy, called query-based selection (QBS), directly queries the smaller caches: when the LLC selects a replacement victim, it queries the smaller caches for approval and uses the information to make its replacement decisions.

With these simple modifications to the LLC replacement policy, Jaleel et al. show that inclusive cache hierarchies can perform similar to non-inclusive hierarchies.

5.5.2   COMPRESSION-AWARE CACHE REPLACEMENT

Increased cache capacity can improve system performance by improving hit rates, but it comes at the cost of area and energy. Compressed caches provide an alternate solution, where data in the cache is compressed to achieve higher effective capacity. For example, if every cache block can be compressed by 4×, the effective cache capacity can be increased 4×.

Of course, not all cache entries can be compressed, so compression generates variable-sized cache blocks, with larger (uncompressed) blocks consuming more cache space than smaller (compressed) blocks. Therefore, compressed caches use a different cache architecture to support variable-sized blocks of different compressibility levels [Sardashti and Wood, 2013, Sardashti et al., 2016], and they need new cache replacement policies to reason about compressibility in addition to temporal locality.

In particular, compressed cache replacement policies must consider the imbalanced benefit of evicting larger blocks vs. evicting smaller blocks. It is beneficial to evict uncompressed blocks because they will generate more free cache capacity, which can be used to cache multiple compressed blocks. But the benefits of evicting an uncompressed block need to balanced with the desire to evict lines that will not be soon re-referenced. We now describe two replacement policies that address these issues.

Compression-Aware Management Policy (CAMP)   CAMP [Pekhimenko et al., 2015] is a replacement policy that takes into account both compressed cache block size and temporal locality. In particular, CAMP computes the *value* of a line by combining information about its size and expected future reuse, and it evicts the line with the lowest value. This component of CAMP is called minimal-value eviction (MVE), and it is based on the observation that it is preferable to evict an uncompressed block with good locality to create room for a set of smaller compressed blocks of the same total size, as long as the set of blocks have enough locality to collectively provides a larger number of hits. Of course, when two blocks have similar locality characteristics, it is preferable to evict the larger cache block.

More concretely, CAMP computes the value of each line as follows: $V_i = p_i/s_i$, where s_i is the compressed block size of block i and p_i is a predictor of locality, such that a larger value of p_i indicates that block i will be re-referenced sooner; p_i is estimated using the RRIP policy [Jaleel et al., 2010b]. Thus, the value increases with higher temporal locality, and it decreases with larger block size. Blocks with a lower overall value are preferable for eviction.

The second component of CAMP, called the size-based insertion policy (SIP), is based on the observation that the compressed size of a block can sometimes be used as an indicator of its reuse characteristics. Thus, s_i can be used to predict p_i. The intuition behind this observation is that elements belonging to the same data structure can have similar compressibility and similar reuse characteristics. SIP exploits this observation by inserting blocks of certain sizes with high priority. To find block sizes that will benefit from high-priority insertion, SIP uses dynamic set sampling [Qureshi et al., 2006].

Base Victim Compression   Gaur et al. [2016] observe that that cache compression and replacement policies can interact antagonistically. In particular, they find that decisions that favor compressed blocks over uncompressed blocks can lead to sub-optimal replacement decisions because they force intelligent replacement policies to change their replacement order. As a result, the resulting compressed cache loses the performance gains from state-of-the-art replacement policies.

To avoid negative interactions with replacement policies, Gaur et al. introduce a cache design that guarantees that all lines that would have existed in an uncompressed cache would also be present in the compressed cache. In particular, their cache design keeps the data array unmodified, but it modifies the tag array to accommodate compression. In particular, the tag array is augmented to associate two tags with each physical way. Logically, the cache is partitioned into a *Baseline* cache, which is managed just like an uncompressed cache, and a *Victim* cache, which opportunistically caches victims from the baseline cache if they can be compressed. This design guarantees a hit rate at least as high as that of an uncompressed cache, so it enjoys the benefits of advanced replacement policies. Furthermore, it can leverage the benefits of compression with simple modifications to the tag array.

5.6 NEW TECHNOLOGY CONSIDERATIONS

For many decades now, caches have been built using SRAM technology, but newer memory technologies promise change as they have been shown to address many limitations of conventional SRAM caches [Fujita et al., 2017, Wong et al., 2016]. An in-depth analysis of these technologies is beyond the scope of this book, but we now briefly discuss design tradeoffs that emerging memory technologies will introduce for cache replacement.

5.6.1 NVM CACHES

Korgaonkar et al. show that last-level caches based on Non-Volatile Memories (NVM) promise high capacity and low power but suffer from performance degradation due to their high write latency [Korgaonkar et al., 2018]. In particular, the high latency of writes puts pressure on the NVM cache's request queues, which puts backpressure on the CPU and interferes with performance-critical read requests.

To mitigate these issues, Korgaonkar et al. propose two cache replacement strategies. First, they introduce a write congestion aware bypass (WCAB) policy that eliminates a large fraction of writes to the NVM cache, while avoiding large reductions in the cache's hit rate. Second, they establish a virtual hybrid cache that absorbs and eliminates redundant writes that would otherwise result in slow NVM writes.

WCAB builds on the observation that traditional bypassing policies [Khan et al., 2010] perform a limited number of write bypasses because they optimize for hit rates instead of write intensity. Unfortunately, naively increasing the intensity of write bypassing adversely affects the cache's hit rate, negating the capacity benefits of NVM caches. Thus, we have a tradeoff between cache hit rate and write intensity. Korgaonkar et al. manage this tradeoff by dynamically estimating write congestion and liveness. If write congestion is high, WCAB sets a high target live score, which means that the liveness score of a line would have to be extremely high for it not to be bypassed. Alternatively, if the write congestion is low, WCAB performs conservative bypassing by reducing the target live score. The liveness score is estimated by sampling a few sets and measuring, for each PC, the fraction of writes that are reused by a subsequent read.

These simple changes to the NVM cache's replacement scheme have a significant impact on its performance and energy, enabling NVM LLCs to utilize the high density of the NVM technology at a performance that is comparable to SRAM caches.

5.6.2   DRAM CACHES

By vertically integrating DRAM and CPU using a high-speed interface, 3D die stacking [Black, 2013, Black et al., 2006] enables dramatic increase in bandwidth between processor and memory. Since the size of stacked-DRAM is not large enough to replace conventional DRAM, researchers have proposed using die-stacked DRAMs as high-bandwidth L4 caches [Chou et al., 2015, Jevdjic et al., 2013, 2014, Jiang et al., 2010, Qureshi and Loh, 2012] (see Figure 5.10).

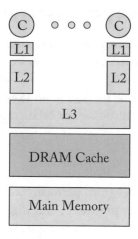

Figure 5.10: Die-stacked DRAM can be used as a large L4 cache; based on [Jiang et al., 2010].

Compared to SRAM caches, DRAM caches offer high capacity—hundreds of megabytes or a few gigabytes—and high bandwidth, but they are much slower, with latencies comparable to DRAM. Many different DRAM cache organizations have been proposed [Jevdjic et al., 2013, Qureshi and Loh, 2012], but a common concern for all DRAM cache replacement policies is to improve hit rates without exacerbating the already high hit latency. For example, some DRAM cache designs co-locate tag with data in the DRAM array [Qureshi and Loh, 2012], and for these caches, filling in data that will not be used adds significant bandwidth overhead due to tag reads and writes [Chou et al., 2015]. For such DRAM cache designs, the bandwidth overhead is large enough to increase hit latency to a point that it is profitable to reduce hit rate at the cost of lower traffic [Chou et al., 2015, Qureshi and Loh, 2012].

Chou et al. propose a Bandwidth-Aware Bypassing scheme to mitigate the harmful effects of adding undesirable data in DRAM caches [Chou et al., 2015]. The key idea is to use set dueling to identify the proportion of the lines that should be bypassed. In particular, two sampling monitors (512K sets each) measure the hit rate of (1) a baseline cache that does not bypass any lines and (2) a cache that probabilistically bypasses 90% of the lines. The baseline cache is likely to have a higher hit rate at the cost of more bandwidth consumption, whereas the probabilistically updated cache will reduce bandwidth consumption at the cost of lower hit rates. Probabilistic bypassing is chosen if its hit rate is not significantly worse than the baseline.

Other DRAM caches are organized at a page granularity so that tags can be stored in fast SRAM arrays [Jevdjic et al., 2013, 2014], but these designs also incur high bandwidth overhead because of the need to fill the entire page on a cache miss. Jiang et al. propose to resolve the bandwidth inefficiency of page-based DRAM caches by only caching *hot pages* (CHOP) in the DRAM cache [Jiang et al., 2010]. To identify hot pages, they use a filter cache (CHOP-FC) that tracks the access frequency of pages that miss in the L3 and only insert a page in the DRAM cache if its access frequency is above a certain threshold (see Figure 5.11). To avoid missing hot pages that get prematurely evicted from the filter cache, they also propose a scheme in which the frequency counters in the filter cache are backed up in main memory (CHOP-MFC). For efficiency, these counters can be added to the page's page-table entry so that they can be retrieved on-chip on a TLB miss. Finally, since the requirements on the DRAM cache vary by workload, they propose adaptive versions of CHOP-FC and CHOP-MFC that turn the filter cache on or off based on memory utilization.

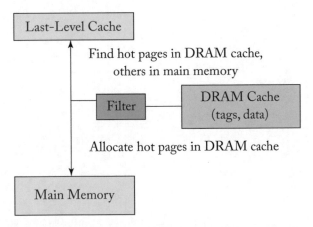

Figure 5.11: CHOP uses a filter cache to identify hot pages; based on [Jiang et al., 2010].

CHAPTER 6

Conclusions

In this book, we have summarized and organized research in cache replacement by defining a new taxonomy of cache replacement policies. While we have not discussed every policy in the literature, we hope that our taxonomy outlines strategic shifts in the design of cache replacement solutions. For example, we have shown that while the first three or four decades of research in cache replacement focused largely on *Coarse-Grained policies*, the last decade has seen a notable shift toward *Fine-Grained policies*.

Cache Replacement Championship 2017 This trend toward Fine-Grained policies is apparent from the results of the 2017 Cache Replacement Championship (CRC), which provides a useful snapshot of the current state-of-the-art.

The CRC is conducted every few years to compare state-of-the-art policies within a uniform evaluation framework. The most recent CRC was held in 2017 in conjunction with ISCA, where submissions were compared on four configurations: (1) a single-core system without a prefetcher, (2) a single-core system with L1/L2 prefetchers, (3) a four-core system without a prefetcher, and (4) a four-core system with L1/L2 prefetchers. Evaluation was done using representative regions of SPEC 2006 benchmarks, and simulations were run for 1 billion instructions after warming up the caches for 200 million instructions.

Figures 6.1[1] and 6.2 summarize the performance of the top-three policies for each configuration. For the two single-core configurations, we include the performance of MIN.[2]

The Hawkeye policy [Jain and Lin, 2016] won the CRC in 2017. However, a more detailed analysis of the CRC results points to some interesting trends. First, while the difference among individual submissions (including those not shown) is small, cache replacement policies have improved significantly, as the top three solutions show impressive improvements over SHiP, which was proposed in 2011. Second, the benefit of intelligent cache replacement is more pronounced on multi-core systems than single-core systems. On the 4-core configuration, the winning solution improves performance by 9.6% in the absence of prefetching (vs. 5% for single-core configuration) and by 7.2% in the presence of prefetching (vs. 2.5% for single-core configuration).

---

[1]LIME is a version of Hawkeye produced by a group other than the original authors.

[2]MIN is provably optimal in the absence of prefetching (first configuration), but not in the presence of prefetching (second configuration). Since MIN requires knowledge of the future, we run the simulation twice to estimate MIN's speedup. The second simulation uses MIN's caching decisions from the first simulation. It is difficult to simulate MIN on multi-core configurations because applications can interleave non-deterministically across different runs.

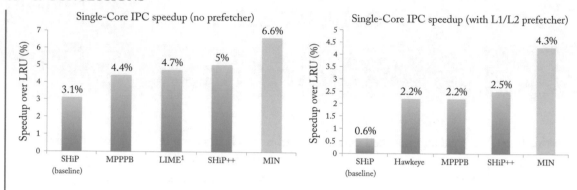

Figure 6.1: Cache replacement championship 2017 single-core results.

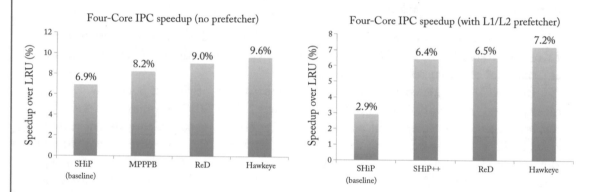

Figure 6.2: Cache replacement championship 2017 multi-core results.

Trends and Challenges   Our comprehesive look at cache replacement policies reveals some clear trends, which point to interesting opportunities for future research.

First, Fine-Grained policies, particularly Classification-based policies, have been shown to outperform traditional Coarse-Grained policies. The most recent Cache Replacement Championship publicized the top three performing finishers of their 15 submissions, and all 3 used Fine-Grained policies. Of course, Fine-Grained policies leverage aging mechanisms from Coarse-Grained policies, so one avenue of future work would be to explore aging schemes customized for Fine-Grained policies.

Second, Fine-Grained policies learn from past behavior, but the key impediment to their performance is their prediction accuracy and their ability to handle inaccurate predictions. Improvements in prediction accuracy are needed to bridge the gap between practical policies and Belady's MIN. As we discuss in Section 4.2.3, state-of-the-art Fine-Grained policies now view cache replacement as a supervised learning problem, which opens up the possibility of applying machine learning to cache replacement.

Third, as seen in Figure 6.1, there remains a considerable gap between state-of-the-art replacement policies and MIN even for single-core configurations, which suggests that there is still room to improve cache replacement for both single-core and multi-core configurations.

Fourth, recent advances have blurred the line between dead block predictors and Fine-Grained cache replacement policies, and we show that even though traditional dead block predictors were designed for a different goal, they fit well in our cache replacement taxonomy.

Finally, there is room to improve cache replacement policies by considering richer factors, including (1) cache replacement policies that cooperate with other components in the system, such as the prefetcher, (2) policies that use performance goals that go beyond improved cache hit rates, and (3) policies that take into account the cache architecture and new memory technologies. Unfortunately, accounting for richer factors within a cache replacement policy remains a challenging problem because of the significantly larger design spaces and because of the lack of efficient optimal algorithms to guide the design process. Nevertheless, with the low-hanging fruit already plucked, these avenues remain promising for future cache replacement research.

Bibliography

2nd Cache Replacement Championship, 2017. http://crc2.ece.tamu.edu/ 34

Jaume Abella, Antonio González, Xavier Vera, and Michael F. P. O'Boyle. IATAC: A smart predictor to turn-off l2 cache lines. *ACM Transactions on Architecture and Code Optimization (TACO)*, 2(1):55–77, 2005. DOI: 10.1145/1061267.1061271 25, 26

Jorge Albericio, Pablo Ibáñez, Víctor Viñals, and José M. Llabería. The reuse cache: Downsizing the shared last-level cache. In *Proc. of the 46th Annual IEEE/ACM International Symposium on Microarchitecture*, pages 310–321, 2013. DOI: 10.1145/2540708.2540735 34

Akhil Arunkumar and Carole-Jean Wu. ReMAP: Reuse and memory access cost aware eviction policy for last level cache management. In *IEEE 32nd International Conference on Computer Design (ICCD)*, pages 110–117, 2014. DOI: 10.1109/iccd.2014.6974670 42

J.-L. Baer and W.-H. Wang. On the inclusion properties for multi-level cache hierarchies. In *Proc. of the 15th Annual International Symposium on Computer Architecture*, pages 73–80, 1988. DOI: 10.1109/isca.1988.5212 53

Nathan Beckmann and Daniel Sanchez. Maximizing cache performance under uncertainty. In *IEEE International Symposium on High Performance Computer Architecture (HPCA)*, pages 109–120, 2017. DOI: 10.1109/hpca.2017.43 5, 25, 35, 36

Laszlo A. Belady. A study of replacement algorithms for a virtual-storage computer. *IBM Systems Journal*, pages 78–101, 1966. DOI: 10.1147/sj.52.0078 31, 35

Bryan Black. Die stacking is happening. In *International Symposium on Microarchitecture*, Davis, CA, 2013. 56

Bryan Black, Murali Annavaram, Ned Brekelbaum, John DeVale, Lei Jiang, Gabriel H. Loh, Don McCaule, Pat Morrow, Donald W. Nelson, Daniel Pantuso, et al. Die stacking (3D) microarchitecture. In *Proc. of the 39th Annual IEEE/ACM International Symposium on Microarchitecture*, pages 469–479, IEEE Computer Society, 2006. DOI: 10.1109/micro.2006.18 56

Burton H. Bloom. Space/time trade-offs in hash coding with allowable errors. *Communications of the ACM*, 13(7):422–426, 1970. DOI: 10.1145/362686.362692 34

Chiachen Chou, Aamer Jaleel, and Moinuddin K. Qureshi. Bear: Techniques for mitigating bandwidth bloat in gigascale dram caches. In *ACM SIGARCH Computer Architecture News*, vol. 43, pages 198–210, 2015. DOI: 10.1145/2749469.2750387 56

Edward Grady Coffman and Peter J. Denning. *Operating Systems Theory*, vol. 973, Prentice Hall, Englewood Cliffs, NJ, 1973. 19

Peter J. Denning. Thrashing: Its causes and prevention. In *Proc. of the Fall Joint Computer Conference, Part I*, pages 915–922, ACM, December 9–11, 1968. DOI: 10.1145/1476589.1476705 9, 10

Nam Duong, Dali Zhao, Taesu Kim, Rosario Cammarota, Mateo Valero, and Alexander V. Veidenbaum. Improving cache management policies using dynamic reuse distances. In *45th Annual IEEE/ACM International Symposium on Microarchitecture (MICRO)*, pages 389–400, 2012. DOI: 10.1109/micro.2012.43 16

Priyank Faldu and Boris Grot. Leeway: Addressing variability in dead-block prediction for last-level caches. In *Proc. of the 26th International Conference on Parallel Architectures and Compilation Techniques*, pages 180–193, 2017. DOI: 10.1109/pact.2017.32 26

Brian Fields, Shai Rubin, and Rastislav Bodík. Focusing processor policies via critical-path prediction. In *Proc. of the 28th Annual International Symposium on Computer Architecture, (ISCA)*, pages 74–85, ACM, 2001. DOI: 10.1145/379240.379253 43, 44

Shinobu Fujita, Hiroki Noguchi, Kazutaka Ikegami, Susumu Takeda, Kumiko Nomura, and Keiko Abe. Novel memory hierarchy with e-STT-MRAM for near-future applications. In *VLSI Design, Automation and Test (VLSI-DAT), International Symposium on*, pages 1–2, IEEE, 2017. DOI: 10.1109/vlsi-dat.2017.7939700 55

Hongliang Gao and Chris Wilkerson. A dueling segmented LRU replacement algorithm with adaptive bypassing. In *JWAC, 1st JILP Workshop on Computer Architecture Competitions: Cache Replacement Championship*, 2010. 13

Jayesh Gaur, Alaa R. Alameldeen, and Sreenivas Subramoney. Base-victim compression: An opportunistic cache compression architecture. In *Computer Architecture (ISCA), ACM/IEEE 43rd Annual International Symposium on*, pages 317–328, 2016. DOI: 10.1109/isca.2016.36 54

Jim Handy. *The Cache Memory Book*. Morgan Kaufmann, 1998. 10

Zhigang Hu, Stefanos Kaxiras, and Margaret Martonosi. Timekeeping in the memory system: Predicting and optimizing memory behavior. In *Computer Architecture, Proceedings. 29th Annual International Symposium on*, pages 209–220, IEEE, 2002. DOI: 10.1145/545214.545239 25, 26

Yasuo Ishii, Mary Inaba, and Kei Hiraki. Access map pattern matching for high performance data cache prefetch. *Journal of Instruction-Level Parallelism*, 13:1–24, 2011. 49

Yasuo Ishii, Mary Inaba, and Kei Hiraki. Unified memory optimizing architecture: Memory subsystem control with a unified predictor. In *Proc. of the 26th ACM International Conference on Supercomputing*, pages 267–278, 2012. DOI: 10.1145/2304576.2304614 49

Akanksha Jain and Calvin Lin. Back to the future: Leveraging belady's algorithm for improved cache replacement. In *Proc. of the International Symposium on Computer Architecture (ISCA)*, June 2016. DOI: 10.1109/isca.2016.17 7, 19, 31, 32, 52, 59

Akanksha Jain and Calvin Lin. Rethinking belady's algorithm to accommodate prefetching. In *ACM/IEEE 45th Annual International Symposium on Computer Architecture (ISCA)*, pages 110–123, 2018. DOI: 10.1109/isca.2018.00020 49, 50, 51, 52

Aamer Jaleel, William Hasenplaugh, Moinuddin Qureshi, Julien Sebot, Simon Steely, Jr., and Joel Emer. Adaptive insertion policies for managing shared caches. In *17th International Conference on Parallel Architectures and Compilation Techniques (PACT)*, pages 208–219, 2008. DOI: 10.1145/1454115.1454145 47

Aamer Jaleel, Eric Borch, Malini Bhandaru, Simon C. Steely Jr., and Joel Emer. Achieving non-inclusive cache performance with inclusive caches: Temporal locality aware (TLA) cache management policies. In *Proc. of the 43rd Annual IEEE/ACM International Symposium on Microarchitecture*, pages 151–162, IEEE Computer Society, 2010a. DOI: 10.1109/micro.2010.52 53

Aamer Jaleel, Kevin B. Theobald, Simon C. Steely Jr., and Joel Emer. High performance cache replacement using re-reference interval prediction (RRIP). In *Proc. of the International Symposium on Computer Architecture (ISCA)*, pages 60–71, 2010b. DOI: 10.1145/1815961.1815971 9, 14, 15, 21, 25, 30, 54

Jaeheon Jeong and Michel Dubois. Cache replacement algorithms with nonuniform miss costs. *IEEE Transactions on Computers*, 55(4):353–365, 2006. DOI: 10.1109/tc.2006.50 39, 40, 41

Djordje Jevdjic, Stavros Volos, and Babak Falsafi. Die-stacked dram caches for servers: Hit ratio, latency, or bandwidth? Have it all with footprint cache. *ACM SIGARCH Computer Architecture News*, 41:404–415, 2013. DOI: 10.1145/2508148.2485957 56, 57

Djordje Jevdjic, Gabriel H. Loh, Cansu Kaynak, and Babak Falsafi. Unison cache: A scalable and effective die-stacked dram cache. In *Proc. of the 47th Annual IEEE/ACM International Symposium on Microarchitecture*, pages 25–37, IEEE Computer Society, 2014. DOI: 10.1109/micro.2014.51 56, 57

Xiaowei Jiang, Niti Madan, Li Zhao, Mike Upton, Ravishankar Iyer, Srihari Makineni, Donald Newell, Yan Solihin, and Rajeev Balasubramonian.   Chop: Adaptive filter-based dram caching for CMP server platforms.   In *HPCA-16 The 16th International Symposium on High-Performance Computer Architecture*, pages 1–12, IEEE, 2010. DOI: 10.1109/hpca.2010.5416642 56, 57

Daniel A. Jiménez. Insertion and promotion for tree-based PseudoLRU last-level caches. In *46th Annual IEEE/ACM International Symposium on Microarchitecture (MICRO)*, pages 284–296, 2013. DOI: 10.1145/2540708.2540733 17, 18

Daniel A. Jiménez and Elvira Teran. Multiperspective reuse prediction. In *Proc. of the 50th Annual IEEE/ACM International Symposium on Microarchitecture (MICRO)*, pages 436–448, 2017. DOI: 10.1145/3123939.3123942 25, 33

Ramakrishna Karedla, J. Spencer Love, and Bradley G. Wherry. Caching strategies to improve disk system performance. *Computer*, (3):38–46, 1994. DOI: 10.1109/2.268884 12, 13

Stefanos Kaxiras, Zhigang Hu, and Margaret Martonosi. Cache decay: Exploiting generational behavior to reduce cache leakage power. In *Computer Architecture. Proc. of the 28th Annual International Symposium on*, pages 240–251, IEEE, 2001. DOI: 10.1109/isca.2001.937453 25

Georgios Keramidas, Pavlos Petoumenos, and Stefanos Kaxiras.   Cache replacement based on reuse-distance prediction. In *25th International Conference on Computer Design (ICCD)*, pages 245–250, 2007. DOI: 10.1109/iccd.2007.4601909 27

Samira Khan, Yingying Tian, and Daniel A. Jiménez.   Sampling dead block prediction for last-level caches. In *43rd Annual IEEE/ACM International Symposium on Microarchitecture (MICRO)*, pages 175–186, 2010. DOI: 10.1109/micro.2010.24 28, 29, 55

Mazen Kharbutli and Rami Sheikh. Lacs: A locality-aware cost-sensitive cache replacement algorithm. *IEEE Transactions on Computers*, 63(8):1975–1987, 2014. DOI: 10.1109/tc.2013.61 42

Mazen Kharbutli and Yan Solihin.   Counter-based cache replacement algorithms.   In *Proc. of the International Conference on Computer Design (ICCD)*, pages 61–68, 2005. DOI: 10.1109/iccd.2005.41 25, 26, 35

Jinchun Kim, Elvira Teran, Paul V. Gratz, Daniel A. Jiménez, Seth H. Pugsley, and Chris Wilkerson. Kill the program counter: Reconstructing program behavior in the processor cache hierarchy. In *Proc. of the 22nd International Conference on Architectural Support for Programming Languages and Operating Systems (ASPLOS)*, pages 737–749, 2017. DOI: 10.1145/3037697.3037701 49

Kunal Korgaonkar, Ishwar Bhati, Huichu Liu, Jayesh Gaur, Sasikanth Manipatruni, Sreenivas Subramoney, Tanay Karnik, Steven Swanson, Ian Young, and Hong Wang. Density tradeoffs of non-volatile memory as a replacement for SRAM based last level cache. In *Proc. of the 45th Annual International Symposium on Computer Architecture*, pages 315–327, IEEE Press, 2018. DOI: 10.1109/isca.2018.00035 55

An-Chow Lai and Babak Falsafi. Selective, accurate, and timely self-invalidation using last-touch prediction. In *The 27th International Symposium on Computer Architecture (ISCA)*, pages 139–148, 2000. DOI: 10.1145/339647.339669 28

An-Chow Lai, Cem Fide, and Babak Falsafi. Dead-block prediction and dead-block correlating prefetchers. In *Proc. of the 28th Annual International Symposium on Computer Architecture (ISCA)*, 2001. DOI: 10.1145/379240.379259 25

D. Lee, J. Choi, J. H. Kim, S. H. Noh, S. L. Min, Y. Cho, and C. S. Kim. LRFU: A spectrum of policies that subsumes the Least Recently Used and Least Frequently Used policies. *IEEE Transactions on Computers*, pages 1352–1361, 2001. DOI: 10.1109/tc.2001.970573 19, 20, 21

Donghee Lee, Jongmoo Choi, Jong-Hun Kim, Sam H. Noh, Sang Lyul Min, Yookun Cho, and Chong Sang Kim. On the existence of a spectrum of policies that subsumes the least recently used (LRU) and least frequently used (LFU) policies. *ACM SIGMETRICS Performance Evaluation Review*, vol. 27, pages 134–143, 1999. DOI: 10.1145/301464.301487 21

Haiming Liu, Michael Ferdman, Jaehyuk Huh, and Doug Burger. Cache bursts: A new approach for eliminating dead blocks and increasing cache efficiency. In *41st Annual IEEE/ACM International Symposium on Microarchitecture (MICRO)*, pages 222–233, 2008. DOI: 10.1109/micro.2008.4771793 26

Madhavan Manivannan, Vassilis Papaefstathiou, Miquel Pericas, and Per Stenstrom. Radar: Runtime-assisted dead region management for last-level caches. In *IEEE International Symposium on High Performance Computer Architecture (HPCA)*, pages 644–656, 2016. DOI: 10.1109/hpca.2016.7446101 48

R. L. Mattson, J. Gegsei, D. R. Slutz, and I. L. Traiger. Evaluation techniques for storage hierarchies. *IBM Systems Journal*, 9(2):78–117, 1970. DOI: 10.1147/sj.92.0078 9

Nimrod Megiddo and Dharmendra S. Modha. ARC: A Self-tuning low overhead replacement cache. In *FAST '03 Proceedings of the 2nd USENIX Conference on File and Storage Technolgies*, 3:115–130, 2003. 22, 23

Anant Vithal Nori, Jayesh Gaur, Siddharth Rai, Sreenivas Subramoney, and Hong Wang. Criticality aware tiered cache hierarchy: A fundamental relook at multi-level cache hierarchies. In

ACM/IEEE 45th Annual International Symposium on Computer Architecture (ISCA), pages 96–109, 2018. DOI: 10.1109/isca.2018.00019 44

Elizabeth J. O'Neil, Patrick E. O'Neil, and Gerhard Weikum. The LRU-K page replacement algorithm for database disk buffering. *ACM SIGMOD Record*, pages 297–306, 1993. DOI: 10.1145/170036.170081 19

Gennady Pekhimenko, Tyler Huberty, Rui Cai, Onur Mutlu, Phillip B. Gibbons, Michael A. Kozuch, and Todd C. Mowry. Exploiting compressed block size as an indicator of future reuse. In *IEEE 21st International Symposium on High Performance Computer Architecture (HPCA)*, pages 51–63, 2015. DOI: 10.1109/hpca.2015.7056021 54

Moinuddin K. Qureshi and Gabe H. Loh. Fundamental latency trade-off in architecting dram caches: Outperforming impractical SRAM-tags with a simple and practical design. In *Proc. of the 45th Annual IEEE/ACM International Symposium on Microarchitecture*, pages 235–246, 2012. DOI: 10.1109/micro.2012.30 56

Moinuddin K. Qureshi and Yale N. Patt. Utility-based cache partitioning: A low-overhead, high-performance, runtime mechanism to partition shared caches. In *The 39th Annual IEEE/ACM International Symposium on Microarchitecture (MICRO)*, pages 423–432, 2006. DOI: 10.1109/micro.2006.49 45, 46, 48

Moinuddin K. Qureshi, Daniel N. Lynch, Onur Mutlu, and Yale N. Patt. A case for MLP-aware cache replacement. In *Proc. of the International Symposium on Computer Architecture (ISCA)*, pages 167–178, 2006. DOI: 10.1145/1150019.1136501 21, 22, 40, 41, 42, 54

Moinuddin K. Qureshi, Aamer Jaleel, Yale N. Patt, Simon C. Steely, and Joel Emer. Adaptive insertion policies for high performance caching. In *Proc. of the International Symposium on Computer Architecture (ISCA)*, pages 381–391, 2007. DOI: 10.1145/1250662.1250709 13, 14, 15, 21, 22, 24, 27, 47

Kaushik Rajan and Ramaswamy Govindarajan. Emulating optimal replacement with a shepherd cache. In *The 40th Annual IEEE/ACM International Symposium on Microarchitecture (MICRO)*, pages 445–454, 2007. DOI: 10.1109/micro.2007.25 17, 18

John T. Robinson and Murthy V. Devarakonda. Data cache management using frequency-based replacement. In *The ACM Conference on Measurement and Modeling Computer Systems (SIGMETRICS)*, pages 134–142, 1990. DOI: 10.1145/98460.98523 19, 20

F. Rosenblatt. *Principles of Neurodynamics: Perceptrons and the Theory of Brain Mechanisms*. Spartan, 1962. DOI: 10.21236/ad0256582 33

Daniel Sanchez and Christos Kozyrakis. Vantage: Scalable and efficient fine-grain cache partitioning. In *Proc. of the International Symposium on Computer Architecture (ISCA)*, pages 57–68, 2011. DOI: 10.1145/2000064.2000073 45

Somayeh Sardashti and David A. Wood. Decoupled compressed cache: Exploiting spatial locality for energy-optimized compressed caching. In *Proc. of the 46th Annual IEEE/ACM International Symposium on Microarchitecture*, pages 62–73, 2013. DOI: 10.1145/2540708.2540715 54

Somayeh Sardashti, André Seznec, and David A. Wood. Yet another compressed cache: A low-cost yet effective compressed cache. *ACM Transactions on Architecture and Code Optimization (TACO)*, 13(3):27, 2016. DOI: 10.1145/2976740 54

Vivek Seshadri, Onur Mutlu, Michael A. Kozuch, and Todd C. Mowry. The evicted-address filter: A unified mechanism to address both cache pollution and thrashing. In *The 21st International Conference on Parallel Architectures and Compilation Techniques (PACT)*, pages 355–366, 2012. DOI: 10.1145/2370816.2370868 33

Vivek Seshadri, Samihan Yedkar, Hongyi Xin, Onur Mutlu, Phillip B. Gibbons, Michael A. Kozuch, and Todd C. Mowry. Mitigating prefetcher-caused pollution using informed caching policies for prefetched blocks. *ACM Transactions on Architecture and Code Optimization (TACO)*, 11(4):51, 2015. DOI: 10.1145/2677956 49

D. Shasha and T. Johnson. 2Q: A low overhead high performance buffer management replacement algorithm. In *Proc. of the 20th International Conference on Very Large Databases*, pages 439–450, Santiago, Chile, 1994. 19

Yannis Smaragdakis, Scott Kaplan, and Paul Wilson. EELRU: Simple and effective adaptive page replacement. *ACM SIGMETRICS Performance Evaluation Review*, pages 122–133, 1999. DOI: 10.1145/301464.301486 11, 12

Santhosh Srinath, Onur Mutlu, Hyesoon Kim, and Yale N. Patt. Feedback directed prefetching: Improving the performance and bandwidth-efficiency of hardware prefetchers. In *Proc. of the 13th International Symposium on High Performance Computer Architecture (HPCA)*, pages 63–74, 2007. DOI: 10.1109/hpca.2007.346185 49

Srikanth T. Srinivasan and Alvin R. Lebeck. Load latency tolerance in dynamically scheduled processors. In *Proc. of the 31st Annual ACM/IEEE International Symposium on Microarchitecture*, pages 148–159, IEEE Computer Society Press, 1998. DOI: 10.21236/ada440304 43, 44

Srikanth T. Srinivasan, R. Dz-Ching Ju, Alvin R. Lebeck, and Chris Wilkerson. Locality vs. criticality. In *Computer Architecture. Proc. of the 28th Annual International Symposium on*, pages 132–143, IEEE, 2001. DOI: 10.1145/379240.379258 43

Lavanya Subramanian, Vivek Seshadri, Arnab Ghosh, Samira Khan, and Onur Mutlu. The application slowdown model: Quantifying and controlling the impact of inter-application interference at shared caches and main memory. In *Proc. of the 48th International Symposium on Microarchitecture*, pages 62–75, ACM, 2015. DOI: 10.1145/2830772.2830803 46

Masamichi Takagi and Kei Hiraki. Inter-reference gap distribution replacement: An improved replacement algorithm for set-associative caches. In *Proc. of the 18th Annual International Conference on Supercomputing*, pages 20–30, ACM, 2004. DOI: 10.1145/1006209.1006213 26

Elvira Teran, Zhe Wang, and Daniel A. Jiménez. Perceptron learning for reuse prediction. In *Microarchitecture (MICRO), 49th Annual IEEE/ACM International Symposium on*, pages 1–12, 2016. DOI: 10.1109/micro.2016.7783705 25, 33, 34

H. S. P. Wong, C. Ahn, J. Cao, H. Y.-Chen, S. B. Eryilmaz, S. W. Fong, J. A. Incorvia, Z. Jiang, H. Li, C. Neumann, K. Okabe, S. Qin, J. Sohn, Y. Wu, S. Yu, X. Zheng, Stanford memory trends, `https://nano.stanford.edu/stanford-memory-trends`, June 6, 2019. 55

Carole-Jean Wu, Aamer Jaleel, Will Hasenplaugh, Margaret Martonosi, Simon C. Steely, Jr., and Joel Emer. SHiP: Signature-based hit predictor for high performance caching. In *44th IEEE/ACM International Symposium on Microarchitecture (MICRO)*, pages 430–441, 2011a. DOI: 10.1145/2155620.2155671 28, 30

Carole-Jean Wu, Aamer Jaleel, Margaret Martonosi, Simon C. Steely, Jr., and Joel Emer. PAC-Man: Prefetch-aware cache management for high performance caching. In *44th Annual IEEE/ACM International Symposium on Microarchitecture (MICRO)*, pages 442–453, 2011b. DOI: 10.1145/2155620.2155672 49, 50

Yuejian Xie and Gabriel H. Loh. PIPP: Promotion/insertion pseudo-partitioning of multi-core shared caches. In *Proc. of the 36th Annual IEEE/ACM International Symposium on Computer Architecture*, pages 174–183, 2009. DOI: 10.1145/1555754.1555778 45, 48

Authors' Biographies

AKANKSHA JAIN

Akanksha Jain is a Research Associate at The University of Texas at Austin. She received her Ph.D. in Computer Science from The University of Texas in August 2016. In 2009, she received B. Tech and M. Tech degrees in Computer Science and Engineering from the Indian Institute of Technology Madras. Her research interests are in computer architecture, with a particular focus on the memory system and on using machine learning techniques to improve the design of memory system optimizations.

CALVIN LIN

Calvin Lin is a University Distinguished Teacher Professor of Computer Science at The University of Texas at Austin. Lin received the BSE in Computer Science from Princeton University in 1985 (Magna Cum Laude) and the Ph.D. in Computer Science from the University of Washington in December 1992. Lin was a postdoc at the University of Washington until 1996, when he joined the faculty at Texas. Lin's research takes a broad view of how compilers and computer hardware can be used to improve system performance, system security, and programmer productivity. He is also Director of UT's Turing Scholars Honors Program, and when he is not working, he can be found chasing his two young sons or coaching the UT men's ultimate frisbee team.

Printed in the United States
by Baker & Taylor Publisher Services